Automating NSX® for vSphere with PowerNSX

The "un-official official" automation tool for NSX

Anthony Burke, Solutions Architect, VMware

With contributions from Nick Bradford and Dale Coghlan

Foreword by Alan Renouf

vmware® PRESS

VMWARE PRESS

Program Managers

Katie Holms
Shinie Shaw

Technical Writer

Rob Greanias

Production Manager

Sappington

Warning & Disclaimer

Every effort has been made to make this book as complete and as accurate as possible, but no warranty or fitness is implied. The information provided is on an "as is" basis. The authors, VMware Press, VMware, and the publisher shall have neither liability nor responsibility to any person or entity with respect to any loss or damages arising from the information contained in this book.

The opinions expressed in this book belong to the author and are not necessarily those of VMware.

VMware, Inc. 3401 Hillview Avenue Palo Alto CA 94304 USA
Tel 877-486-9273 Fax 650-427-5001 www.vmware.com.

Table of Contents

List of Tables

List of Figures

List of Examples

About the Author

Anthony Burke
Solutions Architect,
Network Security Business Unit, VMware

Anthony is a Solution Architect with the Networking & Security Business Unit at VMware. Anthony helps customers transform their networks to support modern network architectures with network virtualization using technologies such as VMware NSX®. His previous experience with emergency services gives him a unique perspective of the network requirements of mission critical environments. Anthony has contributed to and evangelized PowerNSX since its inception. This allows customers to adopt and consume automation of VMware NSX for vSphere® in a familiar and friendly fashion.

About the Reviewers

Nick Bradford
Lead Solutions Architect,
Network Security Business Unit, VMware

With 20 years in a variety of operations, infrastructure, and architecture roles in some of the biggest environments in Australia, Nick has designed and implemented many different infrastructure technologies - including virtualization, storage area networks, datacenter networks, monitoring platforms, and a variety of other supporting technologies. As the author of PowerNSX, he is passionate about automation and its power to enable the full potential of VMware NSX as a part of the VMware SDDC.

Dale Coghlan
Solutions Architect,
Network Security Business Unit, VMware

Dale Coghlan is a solution architect with the Networking & Security Business Unit at VMware. Dale works directly with our NSX customers - from their initial design through to implementation and operationalization of their new environments. Dale has over 17 years of experience in networking and security roles across many verticals and uses that experience to help customers get the best out of the NSX platform.

Acknowledgements

I'd like to thank my family for supporting me with my endeavors. With your support, nothing seems too ambitious or far out of reach, and for that I am eternally grateful. A special shout out to Dale Coghlan and Nick Bradford, my technical contributors. Thank you for your insight, wisdom, nuggets of knowledge, and for cleaning up my horrible grammar. I also want to take the time to thank you for your continuing mentorship.

Thank you also to our VMware NSX for vSphere customers who use PowerNSX. The PowerNSX would not exist as does today without the fanatical support of these users. Finally, a call out to Katie Holms and Shinie Shaw. Thanks for the support in getting this project started and the momentum going. This would not have happened without you both.

Anthony

Audience

This book has been written to cater to administrators and architects of an VMware NSX for vSphere environment. This book should serve as a primer for users with skillsets ranging from beginner to advanced. Some prior knowledge around the following topics will help readers. These topics include:

- Microsoft PowerShell: pipelines, variables, loops, scripting.
- VMware vSphere® PowerCLI™: basic vSphere administration.
- VMware NSX for vSphere: concepts and configuration specifics.

It is assumed that the reader will be familiar with NSX for vSphere constructs and terminology. Additional information about NSX for vSphere can be found in:

- VMware NSX for vSphere Administration guide
- VMware NSX for vSphere Design Guide 3.0

Goal of this book

The goal of this book is to introduce PowerNSX to administrators and architects of an NSX for vSphere environment. This book provides readers insight into the core aspects of PowerNSX. By the end of this book readers will know how to use PowerNSX for:

- Installing and getting started with PowerNSX
- Creating logical switches
- Deploying distributed logical routers (DLR)
- Deploying VMware® NSX Edge™ service gateways (ESG)
- Building NSX Edge services gateway load balancers
- Configuring distributed firewall and objects
- Performing administrative operations
- Using community tools built with PowerNSX

This book will serve as a primer and reference for many day to day tasks performed administrators of VMware NSX for vSphere environments.

Foreword

I will start out this foreword by warning you of the power you now hold in your hands! This book contains fantastic information that any VMware NSX for vSphere engineer would be grateful to digest and use on a daily basis in their journey to provide a fully automated network infrastructure. Congratulations on taking your first step by purchasing this book.

When two strange Australians ask you for a meeting, you don't turn them down for fear of your life or the deadly spiders they may have smuggled with them. In my case I was pleased to meet and work with two of the smartest networking experts I have ever met, abd I am now honored to be writing the foreword as they create this awesome book in front of you.

The amount of work and quality of this work that has been put into PowerNSX has blown my mind on several occasions and provides network administrators a Swiss knife of cmdlets to automate away their day to day troubles and pains, simply produce audit reports and documentation or even automate complex procedures which would have taken days to accomplish in the past.

This book is a must have for any networking engineer, NSX user, automation engineer, VMware full stack administrator, devops engineer or automation title of the month.

Increase your automation skills, progress in your career or just automate your day job away to lead you to higher paid jobs, more time with the family or more time to party.

Go forth and Automate.

Alan Renouf
Sr. Product Line Manager
VMware

Chapter 1

Tools and Cloud Management Platform

Before diving into a book about PowerNSX and how to get started, it is important to distinguish automation tooling from cloud management platforms and detail how each is used.

NOTE
This is not an exhaustive list of tools or cloud management platforms. The commentary and focus is apropos to tooling and those who would use it.

Tooling and Cloud Management Platforms

There is a complimentary relationship between tooling, automation, and cloud management platforms. These concepts may appear to be mutually exclusive or duplicating function, however when considering the consumer - platform operations vs. end user - there is a clear distinction and the complementary nature of these approaches becomes clear.

Many of VMware's most successful customers have embraced end user self-service via a cloud management platform (CMP). They deploy this in conjunction with automation and devops style approaches to platform operations, leveraging tools such as the NSX and vSphere APIs, along with toolsets that consume them like PowerCLI and PowerNSX.

CMPs provide the end user with the resources they need on demand, whereas tooling and automation allow engineers and administrators to efficiently care for and maintain the underlying platform. In some cases, tooling and automation can also be used to interact with the CMP. Although PowerNSX is not concerned with this directly, the extensibility of PowerShell and the availability of third-party extensions (e.g., PowerVRA) make PowerShell an ideal choice of technology for tooling and automation of whole systems. Figure 1.1 highlights this relationship:

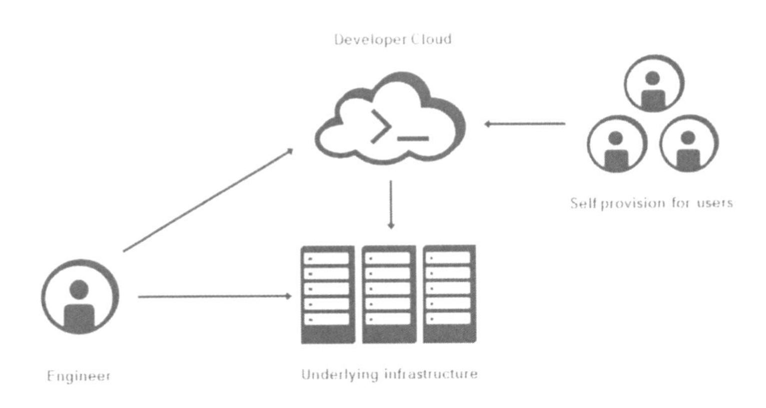

Figure 1.1

Figure 1.1 demonstrates at a very high level the relationship between components; the following section explores each role a little bit more.

- End users are the consumers of the cloud platform. Their focus is on performing job related task such as development, quality assurance, or research. They use a web interface or command line tool to interact with the CMP. One such example is VMware Integrated OpenStack, where the user is provided a web interface, CLI and API interface for their project.
- Engineers are the administrators of the platform. They support the underlying infrastructure that is consumed by the cloud platform on top. Their role is to ensure that there is resource availability and connectivity for all end users. Their tools need to be able to interact with numerous systems and interrogate numerous API endpoints. Engineers use such tools as Ansible, PowerCLI, PowerNSX, Python, and many others.

Cloud Management Platforms ensure that business logic can be applied to resource consumption. Business logic infers that users based on task, role, business unit, or assignment, can request, consume, validate, and destroy resources as seen fit. It also provides the capabilities to perform automated workflows based on request types.

In contrast, tooling provides the means to automate and orchestrate operational and provisional tasks. These include the deployment of new capacity or services, retrieving platform statistics and utilization, and perform wholesale changes across the underlying infrastructure.

Using the right tool for the job is the key here. CMPs and their underlying infrastructures provide API endpoints that are then abstracted by tooling for easy consumption.

Chapter Summary

Tooling provides administrators and engineers the ability to interact with a CMP and the underlying infrastructure at scale. At any stage of the infrastructure's lifecycle, it is possible to use tooling to automate many day to day tasks. The insight provided in this chapter provides the reader the knowledge of where tools such as PowerNSX sit in the world of operations, administration, and cloud management platforms.

Chapter 2

About PowerNSX

What is PowerNSX?

PowerNSX is an open source project primarily developed by members of the solution architect team within VMware's Network and Security Business Unit.

PowerNSX is a PowerShell module that abstracts the VMware NSX API to a set of easily used PowerShell functions.

Since PowerNSX is open source and not developed as an official product, PowerNSX is **not supported** by VMware, and comes with no warranties, express or implied. While there are many customers now using PowerNSX in large scale and complex environments, testing and validation for suitability is recommend before using it in a production environment.

PowerNSX does not cover 100% of the NSX API. It focuses on exposing New, Update, Remove and Get operations for all key NSX functions. It also adds functionality to extend the capabilities of NSX management beyond the native UI or API. With over 250 cmdlets - with more being added constantly - it has significant coverage that addresses most requirements when it comes to automating VMware NSX for vSphere.

Why PowerNSX?

As the usage of PowerNSX increases and discussions take place there is always a question that comes to the forefront. That question is "Why PowerNSX?" Who better to answer that question than the founder and author himself, Nick Bradford?

> "A frequent question we get, is why we chose to develop PowerNSX as a PowerShell module when there were so many languages / environments to choose from, many of them being more 'in vogue' with the devops crowd than PowerShell (I'm looking at you Python guys!).
>
> There are several reasons, but the most relevant one is due to the symbiotic nature of vSphere and NSX, and the massive install base of vSphere customers that VMware already had that were already heavy PowerCLI users.
>
> Very early in the NSX story, we were hearing loud and clear from many of our NSX customers, that because they already had a lot of experience with vSphere, and had been operating it for years using (among other toolsets) PowerCLI, they had the (reasonable) expectation that PowerCLI would naturally be expanded to support NSX as a first-class citizen.
>
> Unfortunately, the reality was that the RESTful NSX API was a completely different beast to the SOAP based vSphere API, and there were several significant technical limitations that prevented extending PowerCLI to support NSX as a formal product.
>
> After several years of waiting for this impasse to be resolved (it wasn't, and while there are plans internally, we are still some way from having formal product support for NSX in PowerCLI), and with the growing realization that our customers, and the VMware field needed more than just 'the API' with which to automate and operate NSX from a CLI and scripting perspective, it was clear to us that an alternative was required.
>
> I have a long history as a vSphere customer myself prior to joining VMware, had utilized PowerCLI heavily, and was aware of its nearly ubiquitous usage within vSphere customers. PowerCLI is an amazing product. It is very mature, and succeeds magnificently in being a toolset for the administrator/engineer rather than being a toolset for the developer, whereas the underlying vSphere API that it abstracts is completely the opposite (read: unfriendly to sysadmins and operations folk).

But why am I talking about PowerCLI and the vSphere API? When you start using the NSX API, you very quickly get confronted with how joined at the hip NSX and vSphere are. You cannot deploy an NSX controller without specifying the datastore and cluster on which it will reside, or which port group it will be connected to. When you define these details, you must do so using the vSphere Managed Object Reference (i.e., MoReF) of the required vSphere entity, a detail that must be determined by interacting with the vSphere API.

Every operation in NSX that relates to a vSphere object (VM, VNIC, port group, Cluster, Resource Pool etc. Think about it, there are a lot of them!), requires a retrieval of that object's MoReF to make the appropriate API call to NSX. Now, to be fair to the Python guys, pyVmomi is a very widely used Python library that abstracts the vSphere API, so why not choose Python? Well, some of our customers already have—and there are open source Python bindings for vSphere as well as Ansible modules now, and for the right customer, this is the right choice—but first and foremost, we could see the need for tooling for our typical vSphere customers and the VMware field—sysadmins and engineers, not developers—who needed to interact heavily with vSphere to be functional using any NSX automation, and already knew and were typically heavy users of PowerCLI.

```
'$ds = Get-Datastore ds001; New-NsxController -
datastore $ds'
```

Look mum. No MoReFs

The choice of platform was easy. PowerNSX was born.

To that end, PowerNSX is designed to work closely with VMware PowerCLI, and PowerCLI users will feel quickly at home using PowerNSX. Together these tools provide a comprehensive command line and automation environment for managing your VMware NSX for vSphere environments.

Remember, PowerNSX is still a work in progress, and it is unlikely that it will ever expose 100% of the NSX API. Feature requests are welcome via the issues tracker on the projects GitHub page."

Where to get PowerNSX

PowerNSX is hosted on VMware's official GitHub repository. The git repository itself is located at https://github.com/vmware/powernsx, and the project wiki can be found at https://powernsx.github.io

Chapter 3

Getting Started with PowerNSX

Installing PowerNSX

Getting started with PowerNSX requires an environment that has PowerShell and PowerCLI installed. PowerCLI Core is now supported, so this can be a Windows, Linux, or macOS device.

The installation steps for the various platforms require the following:

- PowerShell and PowerCLI should be installed for Windows
  - At least PowerShell 3 (recommended 5.1)
  - At time of writing, the Windows installation script will automate the installation of PowerCLI, but this may change shortly with the move of both PowerCLI and PowerNSX to the PowerShell Gallery as a distribution mechanism.
- PowerShell Core and PowerCLI Core have been installed for macOS and Linux
  - At the time of writing, PowerShell Core BETA is *not supported.* The only supported version of PowerShell Core is `v6.0.0-alpha.18`.
  - At the time of writing the supported version of PowerCLI Core is version `1.21`

For more details on supported versions please check https://powernsx.github.io for the latest information.

At time of writing, the recommended method for installing PowerNSX is by running a PowerShell one-liner that downloads and invokes the PowerNSX installer directly from VMware's GitHub repository.

PowerNSX will move its distribution mechanism to PowerShell gallery in the future, so it is recommended to check https://powernsx.github.io for the latest platform-specific installation instructions.

Note
When following the example sections, copy the one-liner from the appropriate installation instructions at https://powernsx.github.io rather than manually typing or copying from this document!

Install PowerNSX on Windows

Example 3.1 outlines the installation process on Windows. The following commands are run from a PowerCLI prompt.

Example 3.1 Installing PowerNSX on Windows

```
PowerCLI C:\> $Branch="master";$url="https://raw.
githubusercontent.com/vmware/powernsx/$Branch/
PowerNSXInstaller.ps1"; try { $wc = new-object Net.
WebClient;$scr = try { $wc.DownloadString($url)} catch
{ if ( $_.exception.innerexception -match "(407)") {
$wc.proxy.credentials = Get-Credential -Message "Proxy
Authentication Required"; $wc.DownloadString($url) }
else { throw $_ }}; $scr | iex } catch { throw $_ }

PowerNSX Installation Tool

PowerNSX is a PowerShell module for VMware NSX (NSX
for vSphere).

PowerNSX requires PowerShell 3.0 or better and VMware
PowerCLI 6.0 or better to function.

This installation script will automatically guide you
through the download and installation of PowerNSX and
its dependencies.  A reboot may be required during the
installation.
```

```
Performing automated installation of PowerNSX.

Continue?

[Y] Yes  [N] No  [?] Help (default is "N"): Y

PowerNSX module is already installed.

Do you want to upgrade to the latest available
PowerNSX?

[Y] Yes  [N] No  [?] Help (default is "N"): Y

PowerNSX installation complete.

PowerNSX requires PowerCLI to function fully!
To get started with PowerNSX, start a new PowerCLI
session.

You can view the cmdlets supported by PowerNSX as
follows:
    get-command -module PowerNSX

You can connect to NSX and vCenter with Connect-
NsxServer.

Head to https://vmware.github.io/powernsx for
documentation, updates and further assistance.

Enjoy!
```

Congratulations! Welcome to using PowerNSX on Windows.

The `master` branch is the main development branch. It is where development between released versions occur. Be warned that whilst testing and validation do occur, the master branch should be considered developmental. A versioned release such as v1 or v2 denote a "point in time" release.

NOTE

If an administrator wants to use a specific branch such as v2, master, or a future released branch then replace the value of `$branch` variable.

Install PowerNSX on macOS

Example 3.2 shows the steps required to install PowerNSX on macOS. The following commands are run from a PowerShell prompt.

Example 3.2 Installing PowerNSX on macOS

```
PS /> $pp = $ProgressPreference;$global:ProgressPrefere
nce = "silentlycontinue";
$Branch="master";$url="https://raw.githubusercontent.
com/vmware/powernsx/$Branch/PowerNSXInstaller.ps1"; try
{ try { $response = Invoke-WebRequest -uri $url; $scr
= $response.content } catch { if ( $_.exception.
innerexception -match "(407)") { $credentials = Get-
Credential -Message "Proxy Authentication Required";
$response = Invoke-WebRequest -uri $url
-proxyCredential $credentials; $scr = $response.content
} else { throw $_ }}; $scr | iex } catch { throw $_
};$global:ProgressPreference = $pp

PowerNSX Installation Tool

PowerNSX is a PowerShell module for VMware NSX (NSX
for vSphere).

PowerNSX requires PowerShell 3.0 or better and VMware
PowerCLI 6.0
or better to function.

This installation script will automatically guide you
through the
download and installation of PowerNSX and its
dependencies.  A reboot
may be required during the installation.

Performing automated installation of PowerNSX.
Continue?
[Y] Yes  [N] No  [?] Help (default is "N"): Y

PowerNSX module not found.
Download and install PowerNSX?
[Y] Yes  [N] No  [?] Help (default is "N"): Y

PowerNSX installation complete.

PowerNSX requires PowerCLI to function fully!
To get started with PowerNSX, start a new PowerCLI
session.
```

```
You can view the cmdlets supported by PowerNSX as
follows:
    get-command -module PowerNSX

You can connect to NSX and vCenter with Connect-
NsxServer.

Head to https://vmware.github.io/powernsx for
documentation,
updates and further assistance.

Enjoy!
```

CAUTION

The `master` branch is the main development branch. It is where development between released versions occur. Be warned that whilst testing and validation do occur, the master branch should be considered developmental. A versioned release such as v1 or v2 denote a "point in time" release.

Congratulations! Welcome to using PowerNSX on macOS.

Install PowerNSX on Linux

Example 3.3 shows the steps required to install PowerNSX on Linux. The following commands are run from a PowerShell prompt.

Example 3.3 Installing PowerNSX on Ubuntu

```
PS /> $pp = $ProgressPreference;$global:ProgressPrefere
nce = "silentlycontinue";
$Branch="master";$url="https://raw.githubusercontent.
com/vmware/powernsx/$Branch/PowerNSXInstaller.ps1"; try
{ try { $response = Invoke-WebRequest -uri $url; $scr
= $response.content } catch { if ( $_.exception.
innerexception -match "(407)") { $credentials = Get-
Credential -Message "Proxy Authentication Required";
$response = Invoke-WebRequest -uri $url
-proxyCredential $credentials; $scr = $response.content
} else { throw $_ }}; $scr | iex } catch { throw $_
};$global:ProgressPreference = $pp

PowerNSX Installation Tool
```

```
PowerNSX is a PowerShell module for VMware NSX (NSX
for vSphere).

PowerNSX requires PowerShell 3.0 or better and VMware
PowerCLI 6.0
or better to function.

This installation script will automatically guide you
through the
download and installation of PowerNSX and its
dependencies.  A reboot
may be required during the installation.

Performing automated installation of PowerNSX.
Continue?
[Y] Yes  [N] No  [?] Help (default is "N"): Y

PowerNSX module not found.
Download and install PowerNSX?
[Y] Yes  [N] No  [?] Help (default is "N"): Y

PowerNSX installation complete.

PowerNSX requires PowerCLI to function fully!
To get started with PowerNSX, start a new PowerCLI
session.

You can view the cmdlets supported by PowerNSX as
follows:
    get-command -module PowerNSX

You can connect to NSX and vCenter with Connect-
NsxServer.

Head to https://vmware.github.io/powernsx for
documentation,
updates and further assistance.

Enjoy!
```

CAUTION
The `master` branch is the main development branch.
It is where development between released versions occur.
Be warned that whilst testing and validation do occur, the master branch should be considered developmental.
A versioned release such as v1 or v2 denote a "point in time" release.

Congratulations! Welcome to using PowerNSX on Linux.

Manual Installation

Administrators who do not have Internet access to install PowerNSX can download the `PowerNSX.psm1` and `PowerNSX.psd1` files. These can be installed by the `Import-Module` cmdlet.

Manual installation requires the `.psd1` and `.psm1` files to be manually downloaded and placed on the filesystem. All prerequisites must be manually met, or the module will fail to load. The `.psd1` and `.psm1` file must be in the same directory. If using Windows, ensure the files are 'unlocked'. It is convention for the module and accompanying manifest to be placed in a directory of the same base name (i.e., PowerNSX)

The recommended locations for Windows users are listed in Table 3.1. These are the default locations defined in the PowerShell variable `$env:PSModulePath`.

Table 3.1 Manual Installation on Windows

Role	Path
All Users	%ProgramFiles%\Common Files\Modules\PowerNSX
Current User	%UserProfile%\Documents\WindowsPowerShell\Modules\PowerNSX

NOTE

If the module is placed in the location defined by the environment variable `$env:PSModulePath` then PowerShell will automatically load it when required, or can be explicitly loaded using the short name (PowerNSX). If an alternate location is chosen an alternate location, use Import-Module and specify the full path to the module manifest (`.psd` file)

Manual installation on Linux and macOS can be performed in a similar fashion. Table 3.2 outlines the directories required.

Table 3.2 Manual Installation on macOS / Linux

Role	Path
All Users	/usr/local/share/powershell/Modules/PowerNSX
Current User	~/.local/share/powershell/Modules/PowerNSX

With the module installed to the default directories it will automatically load when a cmdlet is invoked. Use the command in Example 3.3 to explicitly import the module if desired. As with Windows, macOS and Linux users can explicitly import the module with the short name if it is within the directories specified in `$env:PSModulePath`

Example 3.3 Example 4-5 - Manual Import of module

```
PS /> Import-Module /Users/Administrator/Documents/Git/
powernsx/PowerNSX.psd1
```

After reading this section administrators can manually or automatically install PowerNSX to macOS, Linux, or Windows.

Using the Built-in Help

When building PowerNSX, a lot of attention was given to appropriate and helpful documentation. There are multiple places to go for help when using PowerNSX, but the first place to start is with the native `Get-Help` functionality of PowerShell.

PowerNSX leverages PowerShell's built in documentation framework for all cmdlets. This provides the administrator with the ability to understand a cmdlet and how it is used.

Example 3.4 outlines how to use the `Get-Help` cmdlet with the PowerNSX cmdlet `New-NsxController`.

Example 3.4 Get-Help for New-NsxController

```
PS /> Get-Help New-NsxController

NAME
    New-NsxController
SYNOPSIS
    Deploys a new NSX Controller.
SYNTAX
    New-NsxController [-ControllerName <String>]
[-Confirm] -IpPool <XmlElement> -ResourcePool
<ResourcePoolInterop> -Datastore <DatastoreInterop>
-PortGroup <Object> [-Password <String>] [-Wait]
[-WaitTimeout
    <Int32>] [-Connection <PSObject>]
[<CommonParameters>]
```

```
New-NsxController [-ControllerName <String>] [-Confirm]
-IpPool <XmlElement> -Cluster <ClusterInterop>
-Datastore <DatastoreInterop> -PortGroup <Object>
[-Password <String>] [-Wait] [-WaitTimeout <Int32>] [
    -Connection <PSObject>] [<CommonParameters>]

DESCRIPTION
    An NSX Controller is a member of the NSX
Controller Cluster, and forms the
    highly available distributed control plane for NSX
logical switching and NSX
    Logical Routing.

    The New-NsxController cmdlet deploys a new NSX
Controller.

RELATED LINKS

REMARKS
    To see the examples, type: "get-help New-
NsxController -examples".
    For more information, type: "get-help New-
NsxController -detailed".
    For technical information, type: "get-help New-
NsxController -full".
```

By default, as shown in Example 3.4, the `Get-Help` command shows detail about a cmdlet. The synopsis provides information about the cmdlets function. The syntax will define the parameters accepted by the cmdlet along with the parameter types.

Additionally, most PowerNSX cmdlets also include example code. Example 3.5 shows the output of the `Get-Help New-NsxController -examples` cmdlet.

Example 3.5 Retrieving Examples for a cmdlet

```
PS /> Get-Help New-NsxController -examples

NAME
    New-NsxController
SYNOPSIS
    Deploys a new NSX Controller.
    -------------------------- EXAMPLE 1 --------------
------------

    PS C:\>$ippool = New-NsxIpPool -Name ControllerPool
-Gateway 192.168.0.1 -SubnetPrefixLength 24
-StartAddress 192.168.0.10 -endaddress 192.168.0.20
```

```
$ControllerCluster = Get-Cluster vSphereCluster
    $ControllerDatastore = Get-Datastore
$ControllerDatastoreName -server $Connection.
VIConnection
    $ControllerPortGroup = Get-VDPortGroup
$ControllerPortGroupName -server $Connection.
VIConnection New-NsxController -ipPool $ippool -cluster
$ControllerCluster -datastore $ControllerDatastore
-PortGroup $ControllerPortGroup -password
$DefaultNsxControllerPassword -connection $Connection
-confirm:$false

    -------------------------- EXAMPLE 2 --------------
------------
    PS C:\>$ControllerName = "MyNSXCtrl1"

    $ippool = New-NsxIpPool -Name ControllerPool
-Gateway 192.168.10.1 -SubnetPrefixLength 24
-StartAddress 192.168.10.100 -endaddress 192.168.10.200
    $ControllerCluster = Get-Cluster vSphereCluster
    $ControllerDatastore = Get-Datastore
$ControllerDatastoreName -server $Connection.
VIConnection
    $ControllerPortGroup = Get-VDPortGroup
$ControllerPortGroupName -server $Connection.
VIConnection
    New-NsxController -ControllerName $ControllerName
-ipPool $ippool -cluster $ControllerCluster -datastore
$ControllerDatastore -PortGroup $ControllerPortGroup
-password $DefaultNsxControllerPassword -connection
$Connection -confirm:$false

    -------------------------- EXAMPLE 3 --------------
------------
    PS C:\>A secondary or tertiary controller does not
require a Password to be defined.

    New-NsxController -ipPool $ippool -cluster
$ControllerCluster -datastore $ControllerDatastore
-PortGroup $ControllerPortGroup -connection
$Connection -confirm:$false
```

There are detailed examples for most cmdlets. Example 3.5 above shows three examples on how to deploy NSX controllers with PowerNSX. It outlines examples using various inputs or approaches using a given cmdlet.

If an administrator wants to learn more about the cmdlet the `-detailed` parameter will provide more information about the cmdlets parameters. Below in Example 3.6 the `-detailed` parameter expands further on input parameters for the cmdlet.

Example 3.6 Retrieving Details for a cmdlet

```
PS /> Get-Help New-NsxController -detailed
NAME
    New-NsxController
SYNOPSIS
    Deploys a new NSX Controller.
SYNTAX
    New-NsxController [-ControllerName <String>]
[-Confirm] -IpPool <XmlElement> -ResourcePool
<ResourcePoolInterop> -Datastore <DatastoreInterop>
-PortGroup <Object> [-Password <String>] [-Wait]
[-WaitTimeout
    <Int32>] [-Connection <PSObject>]
[<CommonParameters>]

    New-NsxController [-ControllerName <String>]
[-Confirm] -IpPool <XmlElement> -Cluster
<ClusterInterop> -Datastore <DatastoreInterop>
-PortGroup <Object> [-Password <String>] [-Wait]
[-WaitTimeout <Int32>] [
    -Connection <PSObject>] [<CommonParameters>]

DESCRIPTION
    An NSX Controller is a member of the NSX
Controller Cluster, and forms the
    highly available distributed control plane for NSX
logical switching and NSX
    Logical Routing.

    The New-NsxController cmdlet deploys a new NSX
Controller.
PARAMETERS
    -ControllerName <String>
        Controller Name

    -Confirm [<SwitchParameter>]
        Prompt for confirmation.  Specify as
-confirm:$false to disable confirmation prompt

    -IpPool <XmlElement>
        Pre Created IP Pool object from which
controller IP will be allocated

    -ResourcePool <ResourcePoolInterop>
        vSphere DRS Resource Pool into which to deploy
Controller VM
```

```
-Cluster <ClusterInterop>
        vSphere Cluster into which to deploy the
Controller VM

    -Datastore <DatastoreInterop>
        vSphere Datastore into which to deploy the
Controller VM

    -PortGroup <Object>
        vSphere DVPortGroup OR NSX logical switch
object to connect the Controller VM to.

    -Password <String>

Controller Password (Must be same on all controllers)
    -Wait [<SwitchParameter>]
        Block until Controller Status in API is
'RUNNING' (Will timeout with prompt after -WaitTimeout
seconds)
        Useful if automating the deployment of
multiple controllers (first must be running before
deploying second controller) so you don't have to write
looping code to check status of controller before
continuing.

    -WaitTimeout <Int32>
        Timeout waiting for controller to become
'RUNNING' before user is prompted to continue or
cancel.

    -Connection <PSObject>
        PowerNSX Connection object

    <CommonParameters>

This cmdlet supports the common parameters: Verbose,
Debug,
        ErrorAction, ErrorVariable, WarningAction,
WarningVariable,
        OutBuffer, PipelineVariable, and OutVariable.
For more information, see
        about_CommonParameters (https://go.microsoft.
com/fwlink/?LinkID=113216).
<EXAMPLES SNIPPED FROM OUTPUT>
```

The input parameter information for a given cmdlet provides the administrator with a sense of what objects from the pipeline or variable a given parameter accepts.

The final help option provides all information about a cmdlet. Example 3.7 details using the `-full` parameter with Get-Help on `New-NsxController`.

Example 3.7 Retrieving all Information for a cmdlet

```
PS /> Get-Help New-NsxController -full
NAME
    New-NsxController
SYNOPSIS
    Deploys a new NSX Controller.
SYNTAX
    New-NsxController [-ControllerName <String>]
[-Confirm] -IpPool <XmlElement> -ResourcePool
<ResourcePoolInterop> -Datastore <DatastoreInterop>
-PortGroup <Object> [-Password <String>] [-Wait]
[-WaitTimeout
    <Int32>] [-Connection <PSObject>]
[<CommonParameters>]

    New-NsxController [-ControllerName <String>]
[-Confirm] -IpPool <XmlElement> -Cluster
<ClusterInterop> -Datastore <DatastoreInterop>
-PortGroup <Object> [-Password <String>] [-Wait]
[-WaitTimeout <Int32>] [
    -Connection <PSObject>] [<CommonParameters>]
DESCRIPTION
    An NSX Controller is a member of the NSX
Controller Cluster, and forms the
    highly available distributed control plane for NSX
logical switching and NSX
    Logical Routing.

    The New-NsxController cmdlet deploys a new NSX
Controller.
PARAMETERS
    -ControllerName <String>
        Controller Name
        Required?                    false
        Position?                    named
        Default value
        Accept pipeline input?       false
        Accept wildcard characters?  false

    -Confirm [<SwitchParameter>]
        Prompt for confirmation.  Specify as
-confirm:$false to disable confirmation prompt
        Required?                    false
        Position?                    named
        Default value                True
        Accept pipeline input?       false
        Accept wildcard characters?  false
```

```
-IpPool <XmlElement>
        Pre Created IP Pool object from which
controller IP will be allocated
        Required?                    true
        Position?                    named
        Default value
        Accept pipeline input?       false
        Accept wildcard characters?  false

    -ResourcePool <ResourcePoolInterop>
        vSphere DRS Resource Pool into which to deploy
Controller VM

        Required?                    true
        Position?                    named
        Default value
        Accept pipeline input?       false
        Accept wildcard characters?  false

    -Cluster <ClusterInterop>
        vSphere Cluster into which to deploy the
Controller VM
        Required?                    true
        Position?                    named
        Default value
        Accept pipeline input?       false
        Accept wildcard characters?  false

    -Datastore <DatastoreInterop>
        vSphere Datastore into which to deploy the
Controller VM
        Required?                    true
        Position?                    named
        Default value
        Accept pipeline input?       false
        Accept wildcard characters?  false

    -PortGroup <Object>
        vSphere DVPortGroup OR NSX logical switch
object to connect the Controller VM to
        Required?                    true
        Position?                    named
        Default value
        Accept pipeline input?       false
        Accept wildcard characters?  false
```

```
        -Password <String>
        Controller Password (Must be same on all
controllers)
        Required?                    false
        Position?                    named
        Default value
        Accept pipeline input?       false
        Accept wildcard characters?  false

    -Wait [<SwitchParameter>]
        Block until Controller Status in API is
'RUNNING' (Will timeout with prompt after -WaitTimeout
seconds)
        Useful if automating the deployment of
multiple controllers (first must be running before
deploying second controller)so you don't have to write
looping code to check status of controller before
continuing.
        Required?                    false
        Position?                    named
        Default value                False
        Accept pipeline input?       false
        Accept wildcard characters?  false

    -WaitTimeout <Int32>
        Timeout waiting for controller to become
'RUNNING' before user is prompted to continue or
cancel.
        Required?                    false
        Position?                    named
        Default value                600
        Accept pipeline input?       false
        Accept wildcard characters?  false

    -Connection <PSObject>
        PowerNSX Connection object
        Required?                    false
        Position?                    named
        Default value
$defaultNSXConnection
        Accept pipeline input?       false
        Accept wildcard characters?  false

    <CommonParameters>
        This cmdlet supports the common parameters:
Verbose, Debug,
        ErrorAction, ErrorVariable, WarningAction,
WarningVariable,
        OutBuffer, PipelineVariable, and OutVariable.
For more information, see
        about_CommonParameters (https://go.microsoft.
com/fwlink/?LinkID=113216).
<EXAMPLES SNIPPED FROM OUTPUT>
```

Using the `-full` parameter, an administrator can see the most details about the cmdlets supported parameters. It outlines if a parameter is mandatory, what type of objects are supported from the pipeline, and additional information like default values.

The `Get-Help` cmdlet provides a vast array of help that can be used with any PowerShell cmdlet. It allows for easy access to information about a cmdlet.

Worked Examples and more Usage Information

On top of the included help documentation, considerable effort has gone into documenting common operations on the PowerNSX wiki at https://powernsx.github.io; more contextual assistance is available there. For a fully working example script that leverages PowerCLI and PowerNSX to deploy and configure an NSX environment, see https://github.com/vmware/powernsx/blob/master/Examples/NSXBuildFromScratch.ps1.

NSX Build from Scratch is covered in more detail in Chapter 12.

Chapter Summary

This chapter has introduced how to install PowerNSX across several platforms. It also covered the extensive built-in help functions around each cmdlet. An administrator starting to use PowerNSX should be confident using the automated installer and manual installation methods. Increased familiarity with the `Get-Help` cmdlet will yield tangible benefits when understanding how a cmdlet functions and operates.

The `Get-Help` skills will serve as a foundation moving through the chapters of this book. It will also build developer skills and confidence in creating scripts and tools with PowerNSX.

Chapter 4

Connecting with PowerNSX

PowerNSX communicates with VMware® NSX Manager™ using RESTful API calls. These authenticated API calls are made by establishing a connection to NSX Manager. Chapter 3 demonstrated how to install PowerNSX to an operating system of choice. It is time to get started using PowerNSX. Connecting PowerNSX to NSX Manager is the first step.

Connecting to NSX Manager

Connecting to NSX Manager is the first task a script or administrator will perform when using PowerNSX. By default, PowerNSX connects to NSX Manager and stores the connection information within a default variable. The variable `$DefaultNSXConnection` is used by all cmdlets when performing API calls. Example 4.1 highlights how to connect to NSX Manager.

Example 4.1 Connecting to NSX Manager

```
PS /> Connect-NsxServer -vCenterServer vc-01a.corp.
local
PowerShell credential request
vCenter Server SSO Credentials
User: administrator@vsphere.local
Password for user administrator@vsphere.local: ********

Version                : 6.3.1
BuildNumber            : 5124716
Credential             : System.Management.Automation.
PSCredential
Server                 : nsx-01a.corp.local
Port                   : 443
Protocol               : https
ValidateCertificate : False
VIConnection           : vc-01a.corp.local
DebugLogging           : False
DebugLogfile           : \PowerNSXLog-administrator@
vsphere.local@-2017_07_05_16_26_19.log
```

The `Connect-NsxServer` cmdlet will create the connection to NSX Manager. Furthermore, it will look at the VMware® vCenter™ server registered to NSX Manager. It will establish a connection to both vCenter Server for PowerCLI and NSX Manager for PowerNSX. The cmdlet can use an existing PowerCLI connection if one is already established.

Example 4.2 demonstrates connecting to only an NSX Manager.

Example 4.2 Connecting to NSX Manager

```
PS /> Connect-NsxServer -NsxServer 192.168.103.44
-Username admin -Password VMware1!

PowerNSX requires a PowerCLI connection to the vCenter
server NSX is registered against for proper operation.
Automatically create PowerCLI connection to vc-01a.
corp.local?
[Y] Yes  [N] No  [?] Help (default is "Y"): N

WARNING: Some PowerNSX cmdlets will not be fully
functional without a valid PowerCLI connection to
vCenter server vc-01a.corp.local

Version             : 6.3.1
BuildNumber         : 5124716
Credential          : System.Management.Automation.
PSCredential
Server              : 192.168.103.44
Port                : 443
Protocol            : https
ValidateCertificate : False
VIConnection        :
DebugLogging        : False
DebugLogfile        : \PowerNSXLog-
admin@10.35.255.155-2017_07_05_16_11_51.log
```

There is now an established connection to NSX Manager for use by PowerNSX.

NOTE

The `-DisableVIAutoConnect` parameter will avoid the automatic prompt to connect to vCenter with PowerCLI. Do understand that a PowerCLI connection is required for the correct operation of some PowerNSX cmdlets.

Using Credential Objects

A credential object can be created and passed to specify a username and password in a secure fashion. Example 4.3 demonstrates how to connect using a credential object.

Example 4.3 Connecting to vCenter and NSX Manager with a Credential Object

```
PS /> $cred = Get-Credential
Windows PowerShell credential request
Enter your credentials.
User: administrator@vsphere.local
Password for user administrator@vsphere.local: ********

PS /> Connect-NsxServer -vCenterServer vc-01a.corp.
local -Credential $cred

Version               : 6.3.1
BuildNumber           : 5124716
Credential            : System.Management.Automation.
PSCredential
Server                : nsx-01a.corp.local
Port                  : 443
Protocol              : https
ValidateCertificate : False
VIConnection          : nsx-01a.corp.local
DebugLogging          : False
DebugLogfile          : \PowerNSXLog-administrator@
vsphere.local@-2017_07_05_16_32_47.log
```

After the credential object is created using `Get-Credential` it can be passed using the -credential property to `Connect-NsxServer`. This ensures a more automated connection without prompt or interaction.

Disconnecting from NSX Manager

Where a graceful disconnection of PowerNSX from NSX Manager is required, then Example 4.4 demonstrates how.

Example 4.4 Disconnecting from NSX Manager

```
PS /> Disconnect-NsxServer
```

This will clear the connection and related connection object to NSX Manager.

Overriding the Default Connection

When creating a connection to PowerNSX, it is stored by default in the variable `$DefaultNSXConnection`. This default is used by all cmdlets. There are situations where an administrator may not want to use the default connection. It can be overridden with the `-Connection` parameter. Example 4.5 demonstrates how to use a different connection object.

Example 4.5 Storing and Handling Multiple Connections

```
PS /> $Connection2 = Connect-NsxServer -vCenterServer
10.104.83.100 -Credential $cred2

PS /> $DefaultNSXConnection

Version              : 6.3.1
BuildNumber          : 5124716
Credential           : System.Management.Automation.
PSCredential
Server               : 10.103.82.200
Port                 : 443
Protocol             : https
ValidateCertificate : False
VIConnection         : 10.103.82.100
DebugLogging         : False
DebugLogfile         : \PowerNSXLog-administrator@
vsphere.local@-2017_07_05_15_31_54.log

PS /> $Connection2

Version              : 6.3.1
BuildNumber          : 5124716
Credential           : System.Management.Automation.
PSCredential
Server               : 10.103.83.200
Port                 : 443
Protocol             : https
ValidateCertificate : False
VIConnection         : 10.104.83.100
DebugLogging         : False
DebugLogfile         : \PowerNSXLog-administrator@
vsphere.local@-2017_07_05_15_31_39.log

PS /> Get-NsxIpSet -name RFC1918
```

```
objectId             : ipset-6
objectTypeName       : IPSet
vsmUuid              : 4201B045-B1F9-457F-E621-
B54038A6AFA5
nodeId               : 4b749a6a-bc41-431b-bf24-
cf9e54dcb452
revision             : 1
type                 : type
name                 : RFC1918
description          :
scope                : scope
clientHandle         :
extendedAttributes   :
isUniversal          : false
universalRevision    : 0
inheritanceAllowed   : false
value                : 172.16.0.0/12,10.0.0.0/8,192.168.0.0/16

PS /> Get-NsxIpSet -name RFC1918 -Connection
$Connection2
PS />
```

In Example 4.5 each connection was stored in its respective variable. The contents of each variable display connection specific information about the NSX Manager and vCenter it is connected to.

The cmdlet `Get-NsxIpSet` is used to look up an IP set named `RFC1918`. The first use of `Get-NsxIpSet` uses the `$DefaultNsxConnection`. This returns the IP set named `RFC1918`. Running the same cmdlet using `$Connection2` as the `-Connection` parameter variable returns no result. This is due to the second environment not having an IP set named `RFC1918`.

NOTE
PowerNSX has been written to accommodate numerous NSX environments. As such, each cmdlet has the `-Connection` parameter. This parameter can be passed a specific connection object for that cmdlet. This allows operations to be performed against specific NSX Managers where required.

Chapter Summary

This chapter demonstrates the numerous methods of using PowerNSX to connect to a given environment. Administrators can take advantage of SSO or a direct connection to ensure the right level of access is granted when using the NSX Manager API. With a connection established to vCenter and NSX Manager an administrator can now perform operations with PowerNSX.

Chapter 5

Logical Switching

Logical switching allows VMware NSX for vSphere to provide layer 2 domains that span layer 3 networks. By using a protocol known as Virtual eXtensible Local Area Network (VXLAN), administrators can provision new networks without the requirement for extensive reconfiguration on physical networks.

The goal of this chapter is to provide insight into using PowerNSX for operations pertaining to the use of logical switches.

Working with Transport Zones

Logical switches provide a layer 2 domain for workloads that is independent of the physical topology. These logical switches are deployed within the scope of a transport zone.

These transport zones include clusters of vSphere hosts that make up the boundary of compute that support this logical switch. A host can be a member of more than one transport zone. Depending on the configuration of VMware NSX for vSphere these transport zones can be configured as either local or universal transport zones.

In Example 5.1 the command `Get-NsxTransportZone` will return all configured transport zones - local and universal.

Example 5.1 Retrieving all Transport Zones

```
PS /> Get-NsxTransportZone

objectId              : vdnscope-1
objectTypeName        : VdnScope
vsmUuid               : 564DE9EE-D75F-A2C0-EB82-
843AA89F80E7
nodeId                : de6a0917-606c-432f-9d71-
b96ba2b28706
revision              : 0
type                  : type
name                  : TZ1
description           :
clientHandle          :
extendedAttributes    :
isUniversal           : false
universalRevision     : 0
id                    : vdnscope-1
clusters              : clusters
virtualWireCount      : 0
controlPlaneMode      : UNICAST _ MODE
cdoModeEnabled        : false
objectId              : universalvdnscope
objectTypeName        : VdnScope
vsmUuid               : 564DE9EE-D75F-A2C0-EB82-
843AA89F80E7
nodeId                : de6a0917-606c-432f-9d71-
b96ba2b28706
revision              : 0
type                  : type
name                  : UTZ1
description           :
clientHandle          :
extendedAttributes    :
```

```
isUniversal          : true
universalRevision    : 0
id                   : universalvdnscope
clusters             : clusters
virtualWireCount     : 0
controlPlaneMode     : UNICAST_MODE
cdoModeEnabled       : false
```

The output in Example 5.1 returns two transport zones. One transport zone is configured as local while the other is configured as universal, as shown by the `isUniversal` property. Both are configured for unicast replication as denoted by the `controlPlaneMode` property.

In Example 5.2, a specific transport zone is selected by using the `-Name` parameter. This allows the administrator to return a transport zone by name, as opposed to all transport zones configured as demonstrated in Example 5.1.

Example 5.2 Retrieving a Specific Transport Zone by Name

```
PS /> Get-NsxTransportZone -Name TZ1

objectId             : vdnscope-1
objectTypeName       : VdnScope
vsmUuid              : 564DE9EE-D75F-A2C0-EB82-
843AA89F80E7
nodeId               : de6a0917-606c-432f-9d71-
b96ba2b28706
revision             : 0
type                 : type
name                 : TZ1
description          :
clientHandle         :
extendedAttributes   :
isUniversal          : false
universalRevision    : 0
id                   : vdnscope-1
clusters             : clusters
virtualWireCount     : 0
controlPlaneMode     : UNICAST_MODE
cdoModeEnabled       : false
```

Compared with the output in Example 5.1, the output in Example 5.2 returns a single transport zone object named TZ1. Other parameters such as `-Name`, `-objectId` or `-UniversalOnly` can also be used to filter the result set.

To simplify subsequent commands the next example will store the result in a variable. Example 5.3 demonstrates how to store the output of `Get-NsxTransportZone -Name TZ1` into a variable named `$TZ1`.

Example 5.3 Storing a Transport Zone Object in a Variable

```
PS /> $TZ1 = Get-NsxTransportZone -Name TZ1
PS /> $TZ1

objectId             : vdnscope-1
objectTypeName       : VdnScope
vsmUuid              : 564DE9EE-D75F-A2C0-EB82-
843AA89F80E7
nodeId               : de6a0917-606c-432f-9d71-
b96ba2b28706
revision             : 0
type                 : type
name                 : TZ1
description          :
clientHandle         :
extendedAttributes   :
isUniversal          : false
universalRevision    : 0
id                   : vdnscope-1
clusters             : clusters
virtualWireCount     : 0
controlPlaneMode     : UNICAST_MODE
cdoModeEnabled       : false
```

Issuing `$TZ1` will display the object stored within the variable. When working with cmdlets that accept pipeline input such as `New-NsxLogicalSwitch`, either a cmdlet such as Get-NsxTransportZone, or the variable `$TZ1` can be specified as the first element in the pipeline. The content of `$TZ1` (a PowerNSX transport zone object) is then sent to the input to the next command on the pipeline when a logical switch is created. In this case, this is `New-NsxLogicalSwitch`.

To see which parameters accept pipeline input, review the help documentation for the cmdlet using the `-full` switch. For more details see Chapter 3 - Getting Started with PowerNSX.

Building Logical Switches

The examples so far have been dealing with transport zones which are required to create a logical switch. With the transport zone `TZ1` stored in the variable `$TZ1`, it is now time to create a new logical switch. Example 5-4 highlights the simplicity of creating a new logical switch.

Example 5.4 Creating a New Logical Switch

```
PS /> $TZ1 | New-NsxLogicalSwitch -Name Test-
LogicalSwitch

objectId                 : virtualwire-7
objectTypeName           : VirtualWire
vsmUuid                  : 564DE9EE-D75F-A2C0-EB82-
843AA89F80E7
nodeId                   : de6a0917-606c-432f-9d71-
b96ba2b28706
revision                 : 2
type                     : type
name                     : Test-LogicalSwitch
description              :
clientHandle             :
extendedAttributes       :
isUniversal              : false
universalRevision        : 0
tenantId                 :
vdnScopeId               : vdnscope-1
vdsContextWithBacking    : vdsContextWithBacking
vdnId                    : 5000
guestVlanAllowed         : false
controlPlaneMode         : UNICAST_MODE
ctrlLsUuid               : ef579e88-b34b-4ed5-bdc5-
1b71e22d6987
macLearningEnabled       : false
```

After issuing the command in Example 5.4, a new logical switch called Test-LogicalSwitch has been created. This logical switch has been created in transport zone `TZ1`, which has an object ID of `vdnscope-1`. The `vdnScopeId` property of the logical switch matches the object ID of transport zone TZ1. From here the logical switch can be used by other constructs such as virtual machines, logical routers, and NSX Edges.

NOTE
It is important to understand the behavior of the PowerShell pipeline when performing multi cmdlet operations. Example 5.1 highlighted that two transport zones were already defined. The command `Get-NsxTransportZone | New-NsxLogicalSwitch -name Test-LogicalSwitch` **will create a logical switch for each transport zone passed along the pipeline to the cmdlet** `New-NsxLogicalSwitch`**. This will result in a logical switch named Test-LogicalSwitch being created on** `TZ1` **and** `UTZ1`**. This is one of the strengths of the PowerShell pipeline, but**

it can also be confusing. If in doubt, build pipelines up interactively, piece by piece, verifying the correct results at each step.

By default, a logical switch object inherits the replication mode of the transport zone it is bound to. It is possible to override this with the `-ControlPlaneMode` parameter to the `New-NsxLogicalSwitch` cmdlet. Example 5.5 demonstrates a logical switch with a multicast control plane mode being created within a transport zone configured with a unicast control plane.

Example 5.5 Configuring Replication Mode per Logical Switch

```
PS /> $TZ1 | New-NsxLogicalSwitch -name Test-
LogicalSwitch-MULTICAST -ControlPlaneMode MULTICAST_
MODE

objectId                 : virtualwire-8
objectTypeName           : VirtualWire
vsmUuid                  : 564DE9EE-D75F-A2C0-EB82-
843AA89F80E7
nodeId                   : de6a0917-606c-432f-9d71-
b96ba2b28706
revision                 : 1
type                     : type
name                     : Test-LogicalSwitch-MULTICAST
description              :
clientHandle             :
extendedAttributes       :
isUniversal              : false
universalRevision        : 0
tenantId                 :
vdnScopeId               : vdnscope-1
vdsContextWithBacking    : vdsContextWithBacking
vdnId                    : 5001
multicastAddr            : 239.0.0.0
guestVlanAllowed         : false
controlPlaneMode         : MULTICAST_MODE
macLearningEnabled       : false
```

Building Universal Logical Switches

Building universal logical switches in PowerNSX is done simply by passing a universal transport zone object as an argument for the `-TransportZone` parameter. Example 5.6 demonstrates this using the pipeline.

Example 5.6 Creating a Universal Logical Switch

```
PS /> Get-NsxTransportZone UTZ1 | New-NsxLogicalSwitch
Test-UniversalLogicalSwitch

objectId                  : universalwire-1
objectTypeName            : VirtualWire
vsmUuid                   : 564DE9EE-D75F-A2C0-EB82-
843AA89F80E7
nodeId                    : de6a0917-606c-432f-9d71-
b96ba2b28706
revision                  : 2
type                      : type
name                      : Test-UniversalLogicalSwitch
description               :
clientHandle              :
extendedAttributes        :
isUniversal               : true
universalRevision         : 2
tenantId                  :
vdnScopeId                : universalvdnscope
vdsContextWithBacking     : vdsContextWithBacking
vdnId                     : 6000
guestVlanAllowed          : false
controlPlaneMode          : UNICAST_MODE
ctrlLsUuid                : 3b4c1e30-1161-4b2c-a5fd-
be26fad8b375
macLearningEnabled        : false
```

The `isUniversal` property of the transport zone object passed along the pipeline by the cmdlet `Get-NsxTransportZone` determines the logical switch type. With the transport zone `UTZ1` being of the type universal it will ensure all logical switches bound to it are universal.

Note
When creating a logical switch in Example 5.6 a `-Name` **parameter was not explicitly specified. Most cmdlets in PowerNSX that accept a** `-Name` **parameter, will also accept the first unnamed parameter as the argument to** `-Name`**. In Example 5.6 the argument** `Test-UniversalLogicalSwitch` **is bound to the** `-name` **parameter. This is the only scenario in PowerNSX where unnamed parameters are accepted based on their position.**

Attaching Virtual Machines to Logical Switches

Now that logical switches of all types can be created with PowerNSX, it is time to attach workloads to them. A strength PowerNSX has over other languages, is that it can utilize PowerCLI along the same pipeline. This allows operations like attaching VMs to a logical switch in a few commands.

Example 5.7 shows the steps required to connect a VM to a logical switch.

Example 5.7 Attaching VM1 to Logical Switch Test-UniversalLogicalSwitch

```
PS /> Get-Vm VM1 | Connect-NsxLogicalSwitch -
LogicalSwitch (Get-NsxLogicalSwitch -name Test-
UniversalLogicalSwitch)
```

After retrieving `VM1` with the PowerCLI cmdlet `Get-VM`, it is connected to a logical switch with the PowerNSX cmdlet `Connect-NsxLogicalSwitch`. The logical switch that it is to be connected to is defined with the parameter `-LogicalSwitch`. In this example, the given logical switch is retrieved inline, as opposed to being stored in a variable.

Example 5.7 uses both cmdlets to interact with NSX Manager to perform the network changes for the VMs and cmdlets that retrieve those VMs from vCenter. Example 5.8 demonstrates connecting numerous workloads to a logical switch.

Example 5.8 Attaching Numerous VMs to Logical Switch

```
PS /> $VMs = Get-Vm | Where-Object {$_.name -match
"win-"}
PS /> $Vms

Name                   PowerState Num CPUs MemoryGB
----                   ---------- -------- --------
win-02                 PoweredOff 1        0.500
win-01                 PoweredOff 1        0.500

Connect-NsxLogicalSwitch -VirtualMachine $vms
-LogicalSwitch $ls2
```

The virtual machines are passed to the `Connect-NsxLogicalSwitch` cmdlet via the `-VirtualMachine` parameter. They are attached to the logical switch stored with the `$LS2` variable used by the `-LogicalSwitch` parameter. The cmdlet results in the VMs being connected to the logical switch.

There is a lot going on in this single lined command but when breaking it down into small parts it becomes quite straight forward. The two examples above highlight different methods to achieve the same outcome.

Deleting Logical Switches

Throughout the lifecycle of any network component, there will come a time where it is no longer required and can be removed. PowerNSX provides the ability to remove logical switches in a similar fashion to creating them. Example 5.9 outlines the steps to delete a logical switch.

NOTE
This demonstrates a common approach for workflows in PowerNSX (and PowerCLI and PowerShell more generally). The preferred workflow is 'get a thing or a collection of things, then pass them to the remove- cmdlet to remove them' rather than 'call the remove- cmdlet with a bunch of arguments to define the things to be removed'. This means that typically the remove- cmdlet has a very simple parameter list, with one of its parameters being the PowerNSX object (or collection of objects) representing the thing to be removed. So, the `Get-Widget -color green | Remove-Widget` is the standard approach for deleting objects of any type. This has the benefit of allowing observation of the output of the pipeline for validation purposes up to the remove cmdlet, before adding the final remove- portion.

Example 5.9 Deleting a Logical Switch

```
PS > Get-NsxLogicalSwitch deleteme | Remove-
NsxLogicalSwitch

logical switch removal is permanent.
Proceed with removal of logical switch deleteme?
[Y] Yes  [N] No  [?] Help (default is "N"): Y
PS />
```

This pipeline retrieves the logical switch `deleteme` and pipes it to the `Remove-NsxLogicalSwitch` cmdlet. After the user is prompted for confirmation, it is deleted.

If a logical switch has a virtual machine, DLR, or ESG connected to it, an error is thrown. Example 5.10 highlights this.

Example 5.10 Deleting a Logical Switch with VMs Attached

```
PS /> Get-NsxTransportZone TZ1 | Get-NsxLogicalSwitch
test-logicalswitch | Remove-NsxLogicalSwitch
-confirm:$false
invoke-nsxwebrequest : Invoke-NsxWebRequest : The NSX
API response received indicates a failure. 400 : Bad
Request : Res
ponse Body: <?xml version="1.0" encoding="UTF-8"?>
<error><details>virtualwire-7 resource is still in use
by 3 number of entities.</details><errorCode>849</
errorCode><modu
leName>core-services</moduleName></error>
At /Users/aburke/.local/share/powershell/
Modules/PowerNSX/PowerNSX.psm1:8402 char:21
+ ...     $null = invoke-nsxwebrequest -method "delete"
-uri $URI -connecti ...
+                   ~~~~~~~~~~~~~~~~~~~~~~~~~~~~~~~~~~~~~~
~~~~~~~~~~~~~~~~~~~~
    + CategoryInfo          : InvalidResult: (Invoke-
NsxWebRequest:String) [Invoke-NsxWebRequest],
InternalNsxApiExcept
   ion
    + FullyQualifiedErrorId :
NsxAPIFailureResult,Invoke-NsxWebRequest
```

The error string `virtualwire-7 resource is still in use by 3 number of entities` confirms there are 3 objects still attached. In this case, this is `VM1`, `VM2`, and `VM3` from Example 5.8. They will need to be detached before the cmdlet will delete this logical switch.

WARNING
The parameter `-confirm:$false` will skip the interactive prompt thrown by PowerNSX. The prompt is a guard rail to aid in preventing unexpected changes. Use with caution.

Detaching a VM from a logical switch can be achieved through the `Disconnect-LogicalSwitch` cmdlet. Example 5.11 shows how a set of VMs can be removed from a logical switch before deleting.

NOTE

While the `-VirtualMachine` parameter does accept pipeline input, this demonstrates an alternative method of generating the parameter argument on the fly with the code in parentheses.

Example 5.11 Disconnecting Virtual Machines from a Logical Switch

```
PS /> Disconnect-NsxLogicalSwitch -VirtualMachine
(get-vm | Where-Object {$_.name -match "VM"})

Disconnecting vm3's network adapter from a logical
switch will cause network connectivity loss.
Proceed with disconnection?
[Y] Yes  [N] No  [?] Help (default is "N"): Y
```

Virtual machine `VM3` has now had its vNIC removed from the logical switch.

NOTE

Some VMs have more than a single vNIC. `Disconnect-NsxLogicalSwitch` can define a single NIC through `-NetworkAdapter` parameter. This allows selection of a specific NIC. If all NICs are to be disconnected then the switch `-DisconnectMultipleNics` should be specified.

Progressive Example: Creating Logical Switches

This section provides a progressive example by building logical switches, forming the basis of a logical topology.

The logical switches in Figure 5.1 highlight the switches required for the progressive example. The internal network is a VLAN backed port-group that exists currently on a virtual distributed switch.

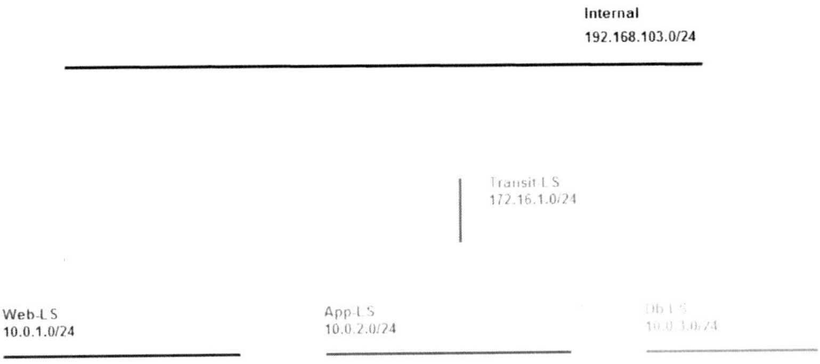

Internal
192.168.103.0/24

Transit-LS
172.16.1.0/24

Web-LS
10.0.1.0/24

App-LS
10.0.2.0/24

Db-LS
10.0.3.0/24

Figure 5.1 New Logical Switches

These logical switches are the foundation of deploying a 3-tier application. In Example 5.13, PowerNSX is used quickly to create four logical switches.

Example 5.12 Creating Logical Switches

```
PS /> $tls = Get-NsxTransportZone TZ1 | New-
NsxLogicalSwitch Transit-LS
PS /> $wls = Get-NsxTransportZone TZ1 | New-
NsxLogicalSwitch Web-LS
PS /> $als = Get-NsxTransportZone TZ1 | New-
NsxLogicalSwitch App-LS
PS /> $dls = Get-NsxTransportZone TZ1 | New-
NsxLogicalSwitch Db-LS
```

No output is generated; the returned objects are stored within a variable for later use.

Example 5.14 validates the logical switches just created for the progressive example.

Example 5.13 Validating Logical Switches

```
PS /> Get-NsxTransportZone TZ1 | Get-NsxLogicalSwitch |
Select-Object Name

Name
----
Transit-LS
Web-LS
App-LS
Db-LS
```

This example demonstrates a simple but common use of the Select-Object cmdlet to filter the property set of the objects in the collection returned by `Get-NsxLogicalSwitch` to include just the `Name` property. This is a human friendly method just getting a list of names for the given objects and can be applied to any object with a name property.

Chapter Summary

This chapter has introduced tasks and operations around logical switching. An administrator using PowerNSX should now be confident in creating and deleting logical switches of various types. Attaching and detaching workloads to these logical switches is now on their tool belt.

These skills will serve as a foundation moving through the chapters in other areas of PowerNSX.

Chapter 6

Distributed Routing

The NSX logical router provides routing and forwarding capability within the hypervisor. This chapter focuses on administration with PowerNSX. As such, an administrator can create, restore, update, and delete logical routers as required. This includes, but is not limited to new routing peers, router interfaces, logical routers, and logical router ancillary configuration.

About Interface Specifications

To flexibly and efficiently support creation of logical routers and NSX Edges with PowerNSX, it was necessary to abstract the operation of defining an interface *specification*. This allows the definition of an arbitrary number of interfaces ahead of time - complete with IP addressing, network connections, etc. - which can then be referenced by variable when using the `New-NsxLogicalRouter` / `New-NsxEdge` cmdlets.

The `New-NsxLogicalRouterInterfaceSpec` and `New-NsxEdgeInterfaceSpec` cmdlets do not talk to the NSX API at all, they just return an XML object that defines a single interface instance for later use by the `New-NsxLogicalRouter` / `New-NsxEdge` cmdlets.

It is also possible to create a logical router or NSX Edge with a single interface, and then iteratively add interfaces using the `New-NsxLogicalRouterInterface` cmdlet which takes address arguments for a single interface instance directly and updates the logical router immediately. This is a much less efficient process though, resulting in multiple round trips to the NSX API, and triggering multiple edge reconfiguration jobs within NSX, which is generally desirable to avoid.

Example 6.1 details an interface specification.

Example 6.1 Interface Spec for Distributed Router

```
PS /> $DLRuplinkspec = New-
NsxLogicalRouterInterfaceSpec -name Uplink -Type
uplink -PrimaryAddress 10.85.100.32 -SubnetPrefixLength
24 -Connected $LS
PS /> $DLRuplinkspec

name            : Uplink
type            : uplink
mtu             : 1500
isConnected     : True
connectedToId   : virtualwire-6
addressGroups   : addressGroups
```

The newly created uplink interface specification provides the inputs needed when creating a new NSX logical router. It is stored in the variable `$DLRuplinkspec` to be passed through to the `-Interface` parameter of `New-NsxLogicalRouter` or `New-NsxLogicalRouterInterface`.

Deploying Logical Routers

With an interface spec created, a new logical router can be created. Example 6.2 demonstrates how to create a new logical router.

Example 6.2 Deploying a new Logical Router

```
PS /Users/aburke> New-NsxLogicalRouter -name DLR
-Interface $dlruplinkspec -Cluster $cl -Datastore $ds
-Tenant "Pepsi" -ManagementPortGroup $ManagementPG

          id                       : edge-12
version              : 2
status               : deployed
tenant               : Pepsi
name                 : DLR
fqdn                 : NSX-edge-12
enableAesni          : true
enableFips           : false
vseLogLevel          : info
appliances           : appliances
cliSettings          : cliSettings
features             : features
autoConfiguration : autoConfiguration
type                 : distributedRouter
isUniversal          : false
mgmtInterface        : mgmtInterface
interfaces           : interfaces
edgeAssistId         : 5000
lrouterUuid          : f82a1962-2d79-4aa5-add6-3a88a746f22f
queryDaemon          : queryDaemon
edgeSummary          : edgeSummary
```

Given that a logical router can have nearly 1000 interfaces configured, the ability to pass an interface spec simplifies the addition of numerous interfaces.

Adding an Interface to an Existing Logical Router

Adding additional networks to a logical router is common practice. A new application or network may be required to be connected to a logical router. The cmdlet `New-NsxLogicalRouterInterface` provides this function. Example 6.3 demonstrates adding a new interface to a pre-deployed logical router.

Example 6.3 Adding a New Interface to a Logical Router

```
PS /> Get-NsxLogicalRouter $lrname | New-
NsxLogicalRouterInterface -Name "NEW-LIF" -ConnectedTo
$LS -PrimaryAddress "10.45.32.1" -SubnetPrefixLength "24"
-Connected -Type internal

interface
---------
interface

PS /> Get-NsxLogicalRouter $lrname | Get-
NsxLogicalRouterInterface -Name NEW-LIF

label             : 138800000010
name              : NEW-LIF
addressGroups     : addressGroups
mtu               : 1500
type              : internal
isConnected       : true
isSharedNetwork   : false
index             : 16
connectedToId     : virtualwire-4
connectedToName   : New-LS
logicalRouterId   : edge-1
```

Once the new interface is created on the logical router, it is reachable per any pre-configured routing settings. If "redistribute connected" is configured, it will ensure the new network is advertised into a given routing protocol. Workloads attached to the network segment will now have gateway connectivity.

Configuring OSPF on Logical Router

Routing is a key component of the logical router. The logical router can support OSPF, BGP, and static routing.

A typical NSX deployment with logical switching, logical routing, and NSX Edge routing will employ the use of dynamic routing between the DLR and NSX Edge, and frequently between the NSX Edge and the physical network to reduce the administrative burden of configuring and maintaining the logical network.

PowerNSX supports the configuration of static routes, as well as OSPF and BGP on logical routers and NSX Edges. This includes BGP neighbor and prefix configuration along with redistribution rules for both protocols.

Example 6.4 configures an existing logical router for OSPF and enables route redistribution.

Example 6.4 Enable OSPF Routing on Logical Router

```
PS /> Get-NsxLogicalRouter DLR | Get-
NsxLogicalRouterRouting | Set-NsxLogicalRouterRouting
-EnableOspf -EnableOspfRouteRedistribution -RouterId
$DlrUplinkPrimaryAddress -ProtocolAddress
$DlrUplinkProtocolAddress -ForwardingAddress
$LdrUplinkPrimaryAddress -confirm:$false

version              : 2
enabled              : true
routingGlobalConfig : routingGlobalConfig
staticRouting        : staticRouting
ospf                 : ospf
logicalrouterId      : edge-6
```

The output of Example 6.4 highlights enabling OSPF with the required parameters such as `-ProtocolAddress` and `-RouterId`.

The parameter `-EnableOspfRouteRedistribution` will redistribute connected interfaces into OSPF. It relies on the default redistribution rule to do this, but the `New-NsxLogicalRouterRedistributionRule` cmdlet will allow manipulation of redistribution rules if required. This will ensure all logical interfaces are automatically advertised via OSPF upon creation. Conversely, it will withdraw the route if the network is no longer present on the router.

Creating a new area for OSPF is performed in Example 6.5 with the `New-NsxLogicalRouterOspfArea` cmdlet.

Example 6.5 Create a New OSPF Area

```
PS />   Get-NsxLogicalRouter DLR | Get-
NsxLogicalRouterRouting | New-NsxLogicalRouterOspfArea
-AreaId $TransitOspfAreaId -Type normal -confirm:$false
areaId type   authentication logicalrouterId
------ ----   -------------- ---------------
0      normal authentication edge-6
```

The area can now be bound to an interface as displayed in Example 6.6.

Example 6.6 Add Logical Router Uplink Interface to OSPF Area

```
PS /> $DlrUplink = Get-NsxLogicalRouter | Get-
NsxLogicalRouterInterface |Where-Object{ $_.name -eq
$TransitLsName }
PS /> Get-NsxLogicalRouter DLR | Get-
NsxLogicalRouterRouting | New-
NsxLogicalRouterOspfInterface -AreaId 0 -vNic
$DlrUplink.index -confirm:$false

vnic            : 2
areaId          : 0
helloInterval   : 10
deadInterval    : 40
priority        : 128
mtuIgnore       : false
logicalrouterId : edge-6
```

With OSPF now configured on the uplink interface of the distributed router, it will attempt to peer with any other routers on the same segment as the uplink interface.

Adding a New Interface

An administrator may need to add an interface to an existing logical router. This can be done by using the `New-NsxLogicalRouterInterface` cmdlet.

NOTE

The observant may note the use of `New-NsxLogicalRouterInterface` rather than `Set-NsxLogicalRouterInterface`, and the use of `Set-NsxEdgeInterface` and not `Add-NsxEdgeInterface`. This is an attempt to remain consistent with the PowerShell semantics around creation of new entities vs. reconfiguring existing ones. The difference here comes about because an NSX Edge always has ten interfaces. They can be connected, disconnected, reconfigured, but not removed or added. In contrast, a logical router has an arbitrary number of interfaces, and by using `New-`

`NsxLogicalRouterInterface`, a new entity is created. The same reasoning is behind the naming of `New-NsxEdgeSubInterface` as opposed to `Set-NsxEdgeSubInterface`, as any Edge trunk interface has an arbitrary number of sub-interfaces.

Example 6.7 demonstrates how to add a new interface to a logical router.

Example 6.7 Adding a New Interface to a Logical Router

```
PS /> Get-NsxLogicalRouter DLR-Coke | New-
NsxLogicalRouterInterface -name Expansion -type
internal -ConnectedTo $ExpansionLS -SubnetPrefixLength
24 -PrimaryAddress 40.0.0.1

interface
---------
interface
```

The logical router `DLR-Coke` just had a new interface added to it called Expansion. When creating a new logical router, it is possible to pass numerous interfaces at once. Example 6.8 shows how.

Example 6.8 Adding Several Interfaces with Interface Specs

```
PS /> $lif1 = New-NsxLogicalRouterInterfaceSpec -name
"Cannery" -PrimaryAddress "31.0.0.1" -SubnetPrefixLength
24 -ConnectedTo $ls1 -Type internal

PS /> $lif2 = New-NsxLogicalRouterInterfaceSpec -name
"Bottle Factor" -PrimaryAddress "32.0.0.1"
-SubnetPrefixLength 24 -ConnectedTo $ls2 -Type internal

PS /> $lif3 = New-NsxLogicalRouterInterfaceSpec -name
"Slab Maker" -PrimaryAddress "33.0.0.1"
-SubnetPrefixLength 24 -ConnectedTo $ls3 -Type internal

PS /> $uplif = New-NsxLogicalRouterInterfaceSpec -name
"Uplink" -PrimaryAddress "30.0.0.1" -SubnetPrefixLength
24 -ConnectedTo $ls0 -Type internal

New-NsxLogicalRouter -name DLR-Coke -Tenant coke
-ManagementPortGroup $pg -Cluster $cl -Datastore $ds
-Interface $lif1, $lif2, $lif3, $uplif
```

The New-NsxLogicalRouterInterfaceSpec cmdlet builds XML documents that can be passed to the -Interface parameter in a consistent fashion. This reduces the number of parameters used when creating a new logical router with numerous interfaces. Example 6.9 reveals how to validate the interface creation.

Example 6.9 Validating Newly Created Interfaces

```
PS /> (Get-NsxLogicalRouter DLR-Coke).interfaces

interface
---------
{Cannery, Bottle Factor, Slab Maker, Uplink}
```

The four interfaces passed to New-NsxLogicalRouterInterfaceSpec during Example 6.8 can be validated by expanding the interfaces property of the logical router DLR-Coke.

Deleting Logical Routers Interfaces

If an interface is no longer required then it can be removed with the Remove-NsxLogicalRouterInterface cmdlet. Example 6.10 demonstrates how to perform this action.

Example 6.10 Removing a Specific Logical Router Interface

```
PS /> Get-NsxLogicalRouter DLR-Coke | Get-
NsxLogicalRouterInterface -name Cannery | Remove-
NsxLogicalRouterInterface

Interface Cannery will be deleted.
Proceed with deletion of interface 10?
[Y] Yes  [N] No  [?] Help (default is "N"): y
PS />
```

The Cannery interface will be removed from the logical router. Any workloads attached to the logical switch associated with the Cannery interface will no longer have a gateway.

Deleting Logical Routers

Removing a logical router can be performed using the Remove-NsxLogicalRouter cmdlet. Example 6.11 demonstrates how.

Example 6.11 Remove a Logical Router

```
PS /> Get-NsxLogicalRouter DLR | Remove-
NsxLogicalRouter

Logical Router removal is permanent.
Proceed with removal of Logical Router DLR?
[Y] Yes  [N] No  [?] Help (default is "N"): Y
```

The logical router has now been deleted.

Progressive Example: Configuring BGP Routing on the Distributed Router

The progressive example will demonstrate the use of PowerNSX to configure BGP to advertise the networks connected to the distributed router to the NSX Edge.

Figure 6.1 provides an overview of the logical topology.

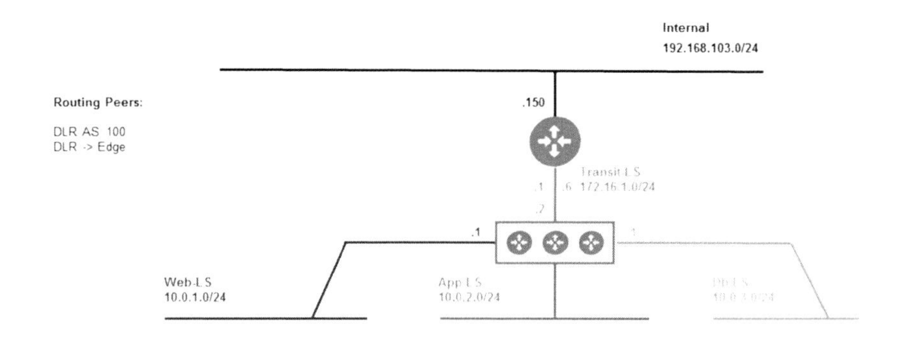

Figure 6.1 BGP Peering on Distributed Router

With a clear picture in mind of what is to be built, it is fine to retrieve the existing logical router and enable BGP. Example 6.12 demonstrates how.

Example 6.12 Configure BGP Routing on Logical Router

```
PS /> Get-NsxLogicalRouter DLR | Get-
NsxLogicalRouterRouting | Set-NsxLogicalRouterRouting
-EnableBgp -ProtocolAddress 172.16.1.3
-ForwardingAddress 172.16.1.2 -LocalAS 100 -RouterId
172.16.1.3 -confirm:$false


version              : 2
enabled              : true
routingGlobalConfig : routingGlobalConfig
staticRouting        : staticRouting
ospf                 : ospf
bgp                  : bgp
logicalrouterId      : edge-9
```

The `-ProtocolAddress` parameter is the IP address assigned to the DLR Control VM. This protocol address is used in the control plane. The `-ForwardingAddress` parameter is the IP address of the logical interface in the data path. The `-RouterID` parameter can be any address. This example chose the same as the `-ProtocolAddress` parameter

Redistribution is needed to ensure BGP learns the correct routing information. It is toggled on a global level. Example 6.13 enables global redistribution for BGP.

Example 6.13 Enable Redistribution for BGP

```
PS /> Get-NsxLogicalRouter DLR | Get-
NsxLogicalRouterRouting | Set-NsxLogicalRouterRouting
-EnableBgpRouteRedistribution -confirm:$false


version              : 3
enabled              : true
routingGlobalConfig : routingGlobalConfig
staticRouting        : staticRouting
ospf                 : ospf
bgp                  : bgp
logicalrouterId      : edge-9
```

With BGP enabled and redistribution enabled, it is time to select what will be redistributed. Connected interfaces are the best option here. This ensures any logical interfaces connected to the DLR are automatically redistributed by BGP. Example 6.14 shows how.

Example 6.14 Redistribute Connected into BGP

```
PS /> Get-NsxLogicalRouter DLR | Get-
NsxLogicalRouterRouting | New-NsxLogicalRouterRedistri
butionRule -FromConnected -Learner bgp -confirm:$false
```

The connected logical interfaces will now be redistributed into BGP as they are added and removed from the distributed logical router.

With the preparation done, the next step is to configure the peering with the upstream NSX Edge. The BGP neighbor requires information about the remote neighbor along with details about the DLR itself. Example 6.15 highlights how.

Example 6.15 Add a BGP Neighbor to the Logical Router

```
PS /> Get-NsxLogicalRouter DLR | Get-
NsxLogicalRouterRouting | New-
NsxLogicalRouterBgpNeighbour -IpAddress 172.16.1.1
-RemoteAS 200 -ForwardingAddress 172.16.1.2
-confirm:$false -ProtocolAddress 172.16.1.3

ipAddress          : 172.16.1.1
protocolAddress    : 172.16.1.3
forwardingAddress  : 172.16.1.2
remoteAS           : 200
remoteASNumber     : 200
weight             : 60
holdDownTimer      : 180
keepAliveTimer     : 60
bgpFilters         :
logicalrouterId    : edge-9
```

The logical router has now been configured to peer with the NSX Edge. Its configuration includes the peers IP address and autonomous system. In the next chapter's progressive example, the NSX Edge will be configured with a neighbor that peers with the logical router to complete the BGP configuration.

Chapter Summary

This chapter has covered key components of the distributed logical router. Some configuration examples related to the NSX distributed logical router have been provided and can be built on for any environment. An administrator using PowerNSX should now be confident in creating NSX DLRs and updating them for routing.

These skills will serve as a foundation moving through the chapters in other areas of PowerNSX.

Chapter 7

NSX Edge Services Gateway

The NSX Edge services gateway is a virtual appliance that provides routing, load balancing, VPN, and additional firewall capabilities within VMware NSX for vSphere. It is commonly used to broker communication from logical switches to VLAN backed port-groups connecting to the existing infrastructure.

The goal of this chapter is to learn about the ESG related functions of PowerNSX. This will allow administrators to perform deployment and operational tasks.

About Interface Specifications

To flexibly and efficiently support creation of logical routers and NSX Edges with PowerNSX, it was necessary to abstract the operation of defining an interface *specification*. This allows the definition of an arbitrary number of interfaces ahead of time - complete with IP addressing, network connection, etc. - which can then be referenced by variable when using the `New-NsxLogicalRouter` / `New-NsxEdge` cmdlets.

The `New-NsxLogicalRouterInterfaceSpec` and `New-NsxEdgeInterfaceSpec` cmdlets do not talk to the NSX API at all, they just return an XML object that defines a single interface instance for later use by the `New-NsxLogicalRouter` / `New-NsxEdge` cmdlets.

It is also possible to create a logical router or NSX Edge with a single interface, and then iteratively add interfaces using the `New-NsxEdgeInterface` cmdlet which takes address arguments for a single interface instance directly and updates the NSX Edge immediately. This is a much less efficient process though, resulting in multiple round trips to the NSX API, and triggering multiple edge reconfiguration jobs within NSX, which is generally desirable to avoid.

Example 7.1 demonstrates the creation of multiple interface specs prior to the creation of the NSX Edge.

Example 7.1 NSX Edge Interface Specifications

```
PS /> $uplink = New-NsxEdgeInterfaceSpec -index 0
-name "Uplink" -Type uplink -Connected $uplinknetwork
-PrimaryAddress "192.168.103.34" -SubnetPrefixLength 24
PS /> $uplink

name                 : Uplink
index                : 0
type                 : uplink
mtu                  : 1500
enableProxyArp       : False
enableSendRedirects  : True
isConnected          : True
portgroupId          : dvportgroup-20
addressGroups        : addressGroups
```

The newly created uplink interface specification provides the inputs needed when creating a new NSX Edge. It is stored in the variable `$uplink` to be passed through to the `-Interface` parameter of `New-NsxEdge` or `New-NsxEdgeInterface`.

Creating a New NSX Edge

Now that an interface specification has been made it is possible to create a new NSX Edge. Example 7.2 builds upon the interface specification created in Example 7.1

Example 7.2 Creating a New NSX Edge

```
PS /> New-NsxEdge -name "Edge" -Datastore $ds -Cluster
$cl -Interface $uplink,$dmz,$internal -Password
"VMware1!VMware1!" -EnableSSH -AutoGenerateRules
-FwDefaultPolicyAllow

id                : edge-5
version           : 1
status            : deployed
tenant            : default
name              : Edge
fqdn              : Edge
enableAesni       : true
enableFips        : false
vseLogLevel       : info
vnics             : vnics
appliances        : appliances
cliSettings       : cliSettings
features          : features
autoConfiguration : autoConfiguration
type              : gatewayServices
isUniversal       : false
hypervisorAssist  : false
queryDaemon       : queryDaemon
edgeSummary       : edgeSummary
```

The new NSX Edge appliance will now be deployed. It can be further configured using PowerNSX for routing, VPN, bridging, firewall, and load balancing functions.

NOTE
Three interfaces are created in this NSX Edge due to using three variables containing different interface specs. This is far more efficient method of defining interfaces if they are known of time during the NSX Edges creation.

NOTE

There are numerous parameters than can be passed when creating an NSX Edge. The parameter `-FwDefaultPolicyAllow` and `-AutoGenerateRules` are useful when establishing connectivity whilst staging an environment.

Adding a New Interface to an Existing NSX Edge

An administrator may need to configure an additional interface on an existing NSX Edge. The cmdlet `Set-NsxEdgeInterface` provides this functionality. Example 7.3 highlights how.

Example 7.3 Adding a New Interface

```
PS /> Get-NsxEdge edge | Get-NsxEdgeInterface -Index 1
| Set-NsxEdgeInterface -Name "Outside World"
-PrimaryAddress "192.168.143.1" -SubnetPrefixLength 24
-Type internal -ConnectedTo $newPG

label                : vNic _ 1
name                 : Outside World
addressGroups        : addressGroups
mtu                  : 1500
type                 : internal
isConnected          : true
index                : 1
portgroupId          : dvportgroup-92
portgroupName        : test _ bridge _ 1
enableProxyArp       : false
enableSendRedirects  : true
edgeId               : edge-3
```

The pipeline has retrieved the given NSX Edge and its interface at `-index 1`. It then has updated the interface per the parameters set for `Set-NsxEdgeInterface`.

In addition to creating interfaces it is possible to pass multiple IP addresses for secondary addresses at the same. This is done by using the `New-NsxAddressSpec` cmdlet. Example 7.4 highlights passing an address spec saved as a variable, as opposed to defining all required parameters.

Example 7.4 Adding Interface Address with an Address Spec

```
$add1 = New-NsxAddressSpec -PrimaryAddress 11.11.11.11
-SubnetPrefixLength 24 -SecondaryAddresses 11.11.11.12,
11.11.11.13
$add2 = New-NsxAddressSpec -PrimaryAddress 22.22.22.22
-SubnetPrefixLength 24 -SecondaryAddresses 22.22.22.23
Get-NsxEdge edge | Get-NsxEdgeInterface -index 5 | Set-
NsxEdgeInterface -ConnectedTo $ls3 -AddressSpec
$add1,$add2 -name "New Interface via Spec" -type
internal


label                : vNic_5
name                 : New Interface via Spec
addressGroups        : addressGroups
mtu                  : 1500
type                 : internal
isConnected          : true
index                : 5
portgroupId          : virtualwire-5
portgroupName        : Internal
enableProxyArp       : false
enableSendRedirects  : true
edgeId               : edge-3
```

The address spec provides a simplified method of creating multiple interfaces in a single command.

Configuring OSPF on NSX Edge

Routing is paramount to dynamic connectivity of networks and their advertisement to other routing domains. The NSX Edge can support OSPF, BGP, and static routing. The following section describes OSPF routing and how PowerNSX configures it.

NOTE
PowerNSX provides support for BGP, OSPF, and static routing on both the NSX Edge and NSX logical router. Not all commands or configurations are included within this booklet. More information can be found using the `Get-Help` **cmdlet in Example 4.6.**

Enabling a routing protocol requires retrieving the desired device and its routing configuration, then enabling it. Example 7.5 demonstrates enabling OSPF on an NSX Edge

Example 7.5 Enable NSX Edge OSPF

```
PS />  Get-NsxEdge Edge | Get-NsxEdgeRouting | Set-
NsxEdgeRouting -EnableOspf -RouterId 192.168.103.34
-confirm:$false


version                 : 4
enabled                 : true
routingGlobalConfig : routingGlobalConfig
staticRouting           : staticRouting
ospf                    : ospf
edgeId                  : edge-3
```

OSPF requires a RouterID to be defined when enabling OSPF. This is passed with the `-RouterID` parameter. The output returned confirms that OSPF is enabled on `edge-3`.

Working with OSPF Areas

By default, new NSX Edges are deployed with an OSPF area of 51. This can be removed with the `Remove-NsxEdgeOspfArea` cmdlet in Example 7.6

Example 7.6 Remove Superfluous Area ID

```
PS />  Get-NsxEdge Edge | Get-NsxEdgeRouting | Get-
NsxEdgeOspfArea -AreaId 51 | Remove-NsxEdgeOspfArea
-confirm:$false
```

Removing this area ensures that only configured areas are included within the OSPF process. With the benign area removed a more suitable area can be made as demonstrated in Example 7.7.

Example 7.7 Adding New OSPF Area

```
PS /> Get-NsxEdge Edge | Get-NsxEdgeRouting | New-
NsxEdgeOspfArea -AreaId 0 -Type normal -confirm:$false

areaId type   authentication edgeId
------ ----   -------------- ------
0      normal authentication edge-3
```

Area 0 has now been created on the given NSX Edge.

OSPF Interface Assignment

The next step is to assign an interface to the area. In Example 7.8 the `New-NsxEdgeOspfInterface` cmdlet will assign a given interface to a specified OSPF area.

Example 7.8 Assigning an NSX Edge Interface to an OSPF Area

```
PS />  Get-NsxEdge Edge | Get-NsxEdgeRouting | New-
NsxEdgeOspfInterface -AreaId 0 -vNic 2 -confirm:$false


vnic          : 2
areaId        : 0
helloInterval : 10
deadInterval  : 40
priority      : 128
mtuIgnore     : false
edgeId        : edge-3
```

Now that an interface has been assigned to an OSPF area, it will try to form a neighbor relationship with other OSPF speakers on the same broadcast domain.

NOTE
The parameter `-vNic` takes the index ID of a given interface to assign it to an OSPF area.

With a single NSX Edge modified for routing, there may be a requirement to modify numerous edges to have a consistent OSPF configuration. These NSX Edges may provide ECMP routing for networks connected south of it via a distributed router. Example 7.9 uses a simple PowerShell pipeline filter on the edge name to return a collection of NSX Edges. This will be used to configure OSPF in one pipeline operation.

Example 7.9 Area and Interface Definition on a Selection of NSX Edges

```
PS /> Get-NsxEdge | Where-Object {$_.Name -match
"ECMP-Edge-"} | Select-Object name, id

name         id
----         --
ECMP-Edge-1  edge-3
ECMP-Edge-2  edge-7
ECMP-Edge-3  edge-8
ECMP-Edge-4  edge-9

PS /> Get-NsxEdge | Where-Object {$_.Name -match
"ECMP-Edge-"} | Get-NsxEdgeRouting | New-
NsxEdgeOspfArea -AreaId 100 -Type normal
-confirm:$false

areaId type   authentication edgeId
------ ----   -------------- ------
100    normal authentication edge-3
100    normal authentication edge-7
100    normal authentication edge-8
100    normal authentication edge-9

PS />  Get-NsxEdge | Where-Object {$_.Name -match
"ECMP-Edge-"} | Get-NsxEdgeRouting | New-
NsxEdgeOspfInterface -AreaId 100 -vNic 0 -confirm:$false

vnic          : 0
areaId        : 100
helloInterval : 10
deadInterval  : 40
priority      : 128
mtuIgnore     : false
edgeId        : edge-3

vnic          : 0
areaId        : 100
helloInterval : 10
deadInterval  : 40
priority      : 128
mtuIgnore     : false
edgeId        : edge-7
```

```
vnic            : 0
areaId          : 100
helloInterval : 10
deadInterval   : 40
priority        : 128
mtuIgnore       : false
edgeId          : edge-8

vnic            : 0
areaId          : 100
helloInterval : 10
deadInterval   : 40
priority        : 128
mtuIgnore       : false
edgeId          : edge-9
```

This example demonstrates the strength of PowerShell and PowerNSX in flexibly applying operations to collections of objects at a time to ensure consistent configuration across all of them.

Progressive Example: Configuring BGP Routing on NSX Edge

The NSX Edge provides connectivity between the logical and physical infrastructure. It requires BGP connectivity between the NSX Edge and the logical router. Figure 7.1 shows the logical topology

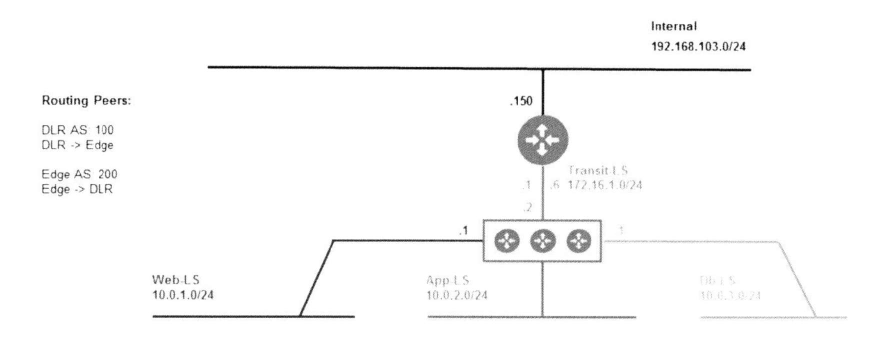

Figure 7.1 Logical Routing Topology

With the topology reviewed the first step is to enable BGP on the NSX Edge. This is performed in Example 7.10.

Example 7.10 Enable BGP on NSX Edge

```
PS /> Get-NsxEdge Edge | Get-NsxEdgeRouting | Set-
NsxEdgeRouting -EnableBgp -RouterId 192.168.103.34
-confirm:$false  -LocalAS 200

version                : 11
enabled                : true
routingGlobalConfig : routingGlobalConfig
staticRouting          : staticRouting
ospf                   : ospf
edgeId                 : edge-3
```

With BGP configured on the NSX Edge, the final step is to configure a BGP neighbor on the NSX Edge. Example 7.11 creates a BGP neighbor with the logical router on the NSX Edge.

Example 7.11 Creating a BGP Peer to the Logical Router

```
PS /> Get-NsxEdge Edge01 | Get-NsxEdgeRouting | New-
NsxEdgeBgpNeighbour -IpAddress 172.16.1.2 -RemoteAS 100

Edge Services Gateway routing update will modify
existing Edge configuration.
Proceed with Update of Edge Services Gateway edge-16?
[Y] Yes  [N] No  [?] Help (default is "N"): Y


ipAddress         : 172.16.1.2
remoteAS          : 100
remoteASNumber : 100
weight            : 60
holdDownTimer   : 180
keepAliveTimer : 60
bgpFilters        :
edgeId            : edge-16
```

The BGP session has been configured on the NSX Edge. It is now peering with the logical router and learning the routes. It will have learned the connected networks on the logical router via their redistribution into BGP on the logical router.

Chapter Summary

This chapter has introduced deployment and configuration tasks related to the NSX Edge. An administrator using PowerNSX should be confident in creating NSX Edges and configuring them for routing and other functions.

These skills will serve as a foundation moving through the chapters in other areas of PowerNSX. The NSX Edge Load Balancer functionality is covered next in Chapter 8.

NSX Edge Load Balancing

VMware NSX for vSphere Edge Services Gateway provide load balancer capability. The NSX Edge load balancing presents administrators with configuration and options to improve service availability for applications.

Enabling the Load Balancer

The NSX load balancer requires three things to be configured correctly. An application profile, a member pool, and a virtual server. These three components form the core requirements of the NSX load balancer.

Example 8.1 Enabling the NSX Load Balancer

```
PS /> Get-NsxEdge vRA7_edge | Get-NsxLoadBalancer |
Set-NsxLoadBalancer -enabled

version                  : 2
enabled                  : true
enableServiceInsertion : false
accelerationEnabled      : false
monitor                  : {default_tcp_monitor,
default_http_monitor, default_https_monitor}
logging                  : logging
edgeId                   : edge-5
```

Now that load balancing is enabled, it can start servicing requests on any configured virtual servers that are defined.

Working with Load Balancer Monitors

With load balancing enabled it is time to create some application specific configuration. Example 8.2 outlines creating a custom monitor for the recently enabled load balancer, and the application it will support.

Example 8.2 Creating a Monitor for a Specific Service

```
PS /> $WebMon = Get-NsxEdge vRA7_edge | Get-
NsxLoadBalancer | New-NsxLoadBalancerMonitor -Name
vRA_Web_HTTPS -TypeHttps -interval
$WebMonitorInterval -timeout $WebMonitorTimeout
-MaxRetries $WebMonitorRetries   -Method
$WebMonitorMethod  -Url $WebMonitorUrl  -receive
$WebMonitorRecieve

PS /> $WebMon
```

```
monitorId   : monitor-4
type        : https
interval    : 3
timeout     : 9
maxRetries  : 3
method      : GET
url         : /wapi/api/status/web
name        : vRA_Web_HTTPS
receive     : REGISTERED
edgeId      : edge-3
```

The custom monitor performs URL validation to ensure a given pool member is active and able to serve valid requests.

NOTE

For the examples simplicity, the values of the parameters have been placed into variables. Each variable has a defined value within a script or the runtime environment.

Example 8.3 further builds on the example now that the pre-requisite monitor for the load balancer pool has been created.

Working with Load Balancer Pools

The pool in Example 8.3 will comprise of specific IP based workloads.

Example 8.3 Creating a New Pool

```
PS /> Get-NsxEdge vRA7_edge | Get-NsxLoadBalancer |
New-NsxLoadBalancerPool -name $WebPoolName
-Description "vRA IaaS Pool" -Transparent:$false
-Algorithm $LbAlgo -Memberspec $webpoolmember1,
$webpoolmember2 -Monitor $Webmon

poolId      : pool-1
name        : vRA-Web-Pool
description : vRA IaaS Pool
algorithm   : round-robin
transparent : false
monitorId   : monitor-4
member      : {vRA-Iaas-01, vRA-Iaas-02}
edgeId      : edge-3
```

A pool is a logical entity that represents members that provide a given service. The values that populates the parameters are specific to the given application.

NOTE
A pool member spec is defined using the `New-NsxLoadBalancerMemberSpec` cmdlet. This cmdlet operates in similar fashion to `New-NsxEdgeInterfaceSpec` where the configuration is created and stored for subsequent use.

Managing Application Profiles

Application profiles represent the type of traffic expected on a virtual server. An application profile is configured per Load Balancer and can be reused across many virtual servers. Example 8.4 creates an application profile.

Example 8.4 Creating an Application Profile

```
PS /> Get-NsxEdge vRA7_edge | Get-NsxLoadBalancer |
New-NsxLoadBalancerApplicationProfile -Name
$WebAppProfileName  -Type $VipProtocol -SslPassthrough

applicationProfileId : applicationProfile-1
name                 : AP-vRA-Web
insertXForwardedFor  : false
sslPassthrough       : true
template             : HTTPS
serverSslEnabled     : false
edgeId               : edge-3
```

Further parameters such as `-sslPassthrough` and `-insertXForwardedFor` can be used here if an application requires it.

Managing Virtual Servers

The virtual server provides the "VIP" to load balance application flows to the pool assigned to it. The virtual server is configured in Example 8.5.

Example 8.5 Building a Virtual Server

```
PS /> Get-NsxEdge vRA7_edge | Get-NsxLoadBalancer |
Add-NsxLoadBalancerVip -name $WebVipName -Description
$WebVipName -ipaddress $EdgeUplinkSecondaryAddress
-Protocol $VipProtocol -Port $HttpPort
-ApplicationProfile $WebAppProfile -DefaultPool $WebPool
-AccelerationEnabled -enabled


version                  : 9
enabled                  : true
enableServiceInsertion   : false
accelerationEnabled      : false
virtualServer            : virtualServer
pool                     : pool
applicationProfile      : applicationProfile
monitor                  : {default_tcp_monitor,
default_http_monitor,
                          default_https_monitor,
vRA_Web_HTTPS}
logging                  : logging
edgeId                   : edge-3
```

The `Add-NsxLoadBalancerVIP` cmdlet takes all previously configured components and enables the virtual server.

NOTE

The monitor property will return all monitors configured on the given NSX Edge. A monitor can be reused across different Virtual Servers on the same NSX Edge. A Virtual Server Pool Monitor is found with `Get-NsxLoadBalancerMonitor` cmdlet and it will list the associated `monitorId`.

A load balancer can have numerous virtual servers. When a virtual server is no longer required it can be removed. Example 8.6 demonstrates removing a specific virtual server.

Example 8.6 Removing a Virtual Server

```
PS /> Get-NsxEdge edge | Get-NsxLoadBalancer | Get-
NsxLoadBalancerVip -name Web_VIP | Remove-
NsxLoadBalancerVip

VIP removal is permanent.
Proceed with removal of VIP virtualServer-1 on Edge
edge-3?
[Y] Yes  [N] No  [?] Help (default is "N"): Y
```

Adding and Removing Pool Members

Workloads sometimes need to be added and removed as an application grows and shrinks. PowerNSX can manipulate pool members. Adding a pool member is shown in Example 8.7.

Example 8.7 Adding a Pool Member

```
PS /> Get-nsxedge Edge01 | Get-NsxLoadBalancer | Get-
NsxLoadBalancerPool WebPool01 | Add-
NsxLoadBalancerPoolMember -name Web-10 -IpAddress
192.168.200.13 -Port 80 -Weight 15 -MinimumConnections
10 -MaximumConnections 3000


poolId      : pool-2
name        : WebPool01
description : WebServer Pool
algorithm   : round-robin
transparent : false
member      : {Web-09, Web-07, Web-01, Web-10}
edgeId      : edge-16
```

With the pool member added it will start receiving traffic based upon the pools configured algorithm. If a pool member needs to be removed it can be done with `Remove-NsxLoadBalancerPoolMember` as shown in Example 8.8.

Example 8.8 Removing a Pool Member

```
PS /> Get-nsxedge Edge01 | Get-NsxLoadBalancer | Get-
NsxLoadBalancerPool WebPool01 | Get-
NsxLoadBalancerPoolMember Web-01 | Remove-
NsxLoadBalancerPoolMember

Pool Member removal is permanent.
Proceed with removal of Pool Member member-1?
[Y] Yes  [N] No  [?] Help (default is "N"): y


poolId      : pool-2
name        : WebPool01
description : WebServer Pool
algorithm   : round-robin
transparent : false
member      : member
edgeId      : edge-16
```

Managing Application Rules

Application rules allow further actions to happen based on traffic being load balanced. Configuring an application rule is demonstrated in Example 8.9.

Example 8.9 Creating an Application Rule

```
PS C:\>Get-NsxEdge Edge01 | Get-NsxLoadBalancer | New-
NsxLoadBalancerApplicationRule -name AR-Redirect-vC
-script $script
```

This creates an application rule for use by a virtual server.

Progressive Example: Configuring Load Balancing Web and App Tiers

With the logical topology configured with routing and logical switches it is time to prepare the application side of things. NSX Edge load balancers provide many functions suitable to load balance most enterprise applications. To visualize the goal of this example, Figure 8.1 depicts the logical topology

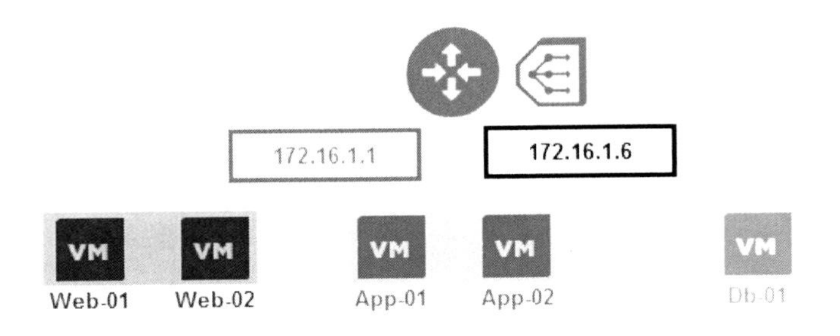

Figure 8.1 Load Balancing Logical Topology

The load balancer will need to be configured like this to support the progressive example.

The first step will enable the NSX Edge load balancer. This will use the existing edge previously deployed. Example 8.10 enables the load balancer.

Example 8.10 Enabling the Load Balancer

```
PS /> Get-NsxEdge Edge | Get-NsxLoadBalancer | Set-
NsxLoadBalancer -Enabled

version                    : 13
enabled                    : true
enableServiceInsertion     : false
accelerationEnabled        : false
virtualServer              : virtualServer
pool                       : pool
applicationProfile         : applicationProfile
monitor                    : {default_tcp_monitor,
default_http_monitor, default_https_monitor, vRA_
Web_HTTPS}
logging                    : logging
edgeId                     : edge-3
```

With the load balancer enabled the configuration for the progressive example can be created. Before creating a virtual server, several required components need to be built. Example 8.11 creates the monitors needed for the application.

Example 8.11 Creating Monitors

```
PS /> $WebMonitor =  Get-NsxEdge Edge | Get-
NsxLoadBalancer | Get-NsxLoadBalancerMonitor -Name
$WebMonitorName

PS /> $AppMonitor =  Get-NsxEdge Edge | Get-
NsxLoadBalancer | Get-NsxLoadBalancerMonitor -Name
$AppMonitorName
```

The newly created monitors have been stored into a variable for later use. These monitors will be used when building pools in Example 8.12.

Example 8.12 Building and Populating the Pools

```
PS /> $WebPool = Get-NsxEdge Edge | Get-
NsxLoadBalancer | New-NsxLoadBalancerPool -name
Web-Pool -Description "Web Tier Pool"
-Transparent:$false -Algorithm round-robin -Memberspec
$webpoolmember1, $webpoolmember2 -Monitor $WebMonitor

PS /> $null = $WebPool | Add-NsxLoadBalancerPoolMember
-name "Web-01" -IpAddress 10.0.1.10 -Port $HttpPort
PS /> $null = $WebPool | Add-NsxLoadBalancerPoolMember
-name "Web-02" -IpAddress 10.0.1.11 -Port $HttpPort

PS /> $AppPool = Get-NsxEdge Edge | Get-NsxLoadBalancer
| New-NsxLoadBalancerPool -name App-Pool -Description
"App Tier Pool" -Transparent:$false -Algorithm round-
robin -Monitor $AppMonitor

PS /> $null = $AppPool | Add-NsxLoadBalancerPoolMember
-name "App-01" -IpAddress 10.0.2.10 -Port $HttpPort
PS /> $null = $AppPool | Add-NsxLoadBalancerPoolMember
-name "App-02" -IpAddress 10.0.2.11 -Port $HttpPort
```

Two load balancer pools have been created and they have members assigned to their respective pools.

NOTE
Using `$null` or piping a cmdlet to `Out-Null` will delete output instead of returning it to the console. This allows for cleaner output when executing scripts.

Application profiles define what traffic is expected on a virtual server.

Example 8.13 Creating Application Profiles

```
PS /> $WebAppProfile = Get-NsxEdge Edge | Get-
NsxLoadBalancer | New-NsxLoadBalancerApplicationProfile
-Name "AP-WebTier" -Type $VipProtocol
PS /> $AppAppProfile = Get-NsxEdge $EdgeName | Get-
NsxLoadBalancer | new-NsxLoadBalancerApplicationProfile
-Name "AP-AppTier" -Type $VipProtocol
```

Both virtual servers require an application profile and have been stored in variables. These will be used when creating the virtual servers. The final part of this progressive example culminates in the creation of the virtual servers, or VIPs in Example 8.14.

Example 8.14 Constructing the Virtual Servers

```
PS /> $null = Get-NsxEdge Edge | Get-NsxLoadBalancer |
Add-NsxLoadBalancerVip -name VIP-Web -ipaddress
192.168.103.150 -Protocol $VipProtocol -Port 80
-ApplicationProfile $WebAppProfile -DefaultPool $WebPool
-AccelerationEnabled

PS /> $null = Get-NsxEdge Edge | Get-NsxLoadBalancer |
Add-NsxLoadBalancerVip -name VIP-App -ipaddress
172.16.1.6 -Protocol $VipProtocol -Port 80
-ApplicationProfile $WebAppProfile -DefaultPool $WebPool
-AccelerationEnabled | out-null
```

The virtual servers have been created. They now will perform load balancing based on incoming traffic to the virtual server. It will then distribute the traffic based on the configuration to the respective pool members.

Chapter Summary

This chapter has introduced NSX load balancing functions. An administrator using PowerNSX should be confident in leveraging PowerShell to retrieve, create, and manage application load balancers. Furthermore, providing repeatable configuration at large or quickly delivering additional function can be achieved with PowerNSX.

These skills will serve not only in managing an VMware NSX for vSphere environment, but the infrastructure as well.

Chapter 9

Distributed Firewall and Objects

VMware NSX for vSphere delivers a capability known as distributed firewall (DFW). The DFW provides kernel level, stateful firewall function at the Virtual NIC. This unique filtering point provides near line rate network protection.

The goal of this chapter is to provide insight into using PowerNSX for operations pertaining to the use of the DFW and its supporting objects.

Working with Firewall Sections

Firewall sections allow administrators to group similar DFW rules together.

The command `New-NsxFirewallSection` Example 9.1 creates a new firewall section.

Example 9.1 New Firewall Section

```
PS /> New-NsxFirewallSection -name New-Application

id               : 1889
name             : New-Application
generationNumber : 1495794592418
timestamp        : 1495794592418
type             : LAYER3
```

Firewall sections can be retrieved with `Get-NsxFirewallSection`. When creating firewall rules, a firewall section is required to be passed along the pipeline to `New-NsxFirewallRule`. This is the same behavior exhibited by `New-NsxLogicalSwitch`, that requires `Get-NsxTransportZone`. Example 9.2 retrieves the firewalls section by name.

Example 9.2 Retrieving a Firewall Section

```
PS /> Get-NsxFirewallSection -name New-Application

id               : 1889
name             : New-Application
generationNumber : 1495794592418
timestamp        : 1495794592418
type             : LAYER3
```

Like most commands in PowerNSX `Get-NsxFirewallSection` provides the ability to search on name.

NOTE
From a DFW perspective, there are three types of firewall sections. They are `-layer2sections`, `-layer3redirectsections` and `-layer3sections`. These are parameters that can be defined when using `Get-NsxFirewallSection` and `New-NsxFirewallSection`.

Creating Firewall Rules

Creating new DFW rules can be performed with PowerNSX. This allows administrators to start protecting applications and infrastructure.

Example 9.3 outlines the basic rule created on the DFW

Example 9.3 Creating a Basic DFW Rule

```
PS /> Get-NsxFirewallSection New-Application | New-
NsxFirewallRule -name Example-1 -Action Allow

id             : 2008
disabled       : false
logged         : false
name           : Example-1
action         : allow
appliedToList  : appliedToList
sectionId      : 1889
direction      : inout
packetType     : any
```

The new DFW rule is created in the firewall section called New-Application. This rule has defined the bare minimum required for a new DFW rule. When no source, destination or service is defined, those properties in the resultant DFW rule will be "any". As such, this rule will permit any traffic on any port and protocol to any destination.

Removing Firewall Rules

Figure 9.4 removes the rule that was just created.

Example 9.4 Removing a Firewall Rule

```
PS /> Get-NsxFirewallSection New-Application | Get-
NsxFirewallRule -name Example-1 | Remove-
NsxFirewallRule

Firewall Rule removal is permanent and cannot be
reversed.
Proceed with removal of Rule Example-1?
[Y] Yes  [N] No  [?] Help (default is "N"): Y
```

The user is prompted to remove the DFW rule. This prompt ensures a user does not accidently remove a DFW rule. The parameter `-confirm` with the option `$false` can be used to override this.

It is possible to remove only the DFW rules within a section without removing the DFW section itself.

Example 9.5 Removing all Rules within a Section

```
PS /> Get-NsxFirewallSection New-Application | Get-
NsxFirewallRule | Remove-NsxFirewallRule
-Confirm:$false

PS /> Get-NsxFirewallSection New-Application | Get-
NsxFirewallRule
```

The output in Example 9.5 will retrieve the NSX firewall section `New-Application` and its associated rules through `Get-NsxFirewallRule`. All rules within that section will be removed with `Remove-NsxFirewallRule`. The parameter `-Confirm:$false` will not prompt the administrator.

Now that the permissive and non-secure rules have been removed, it is time to create more targeted rules using defined sources and destinations. This can be through IP addresses, vCenter/NSX objects, or NSX security objects.

Creating Objects for use in Distributed Firewall Rules

VMware NSX for vSphere can use a mix of vCenter and NSX objects as part of DFW rules. These objects can be the source, destination, or the applied to object for DFW rules.

Example 9.6 highlights creating a new IPSet

Example 9.6 Creating a New IPSet

```
PS /> New-NsxIpSet -Name IPS-RFC1918 -Description
"RFC1918 subnets" -IPAddresses "10.0.0.0/8,172.16.0.0/12"

objectId          : ipset-2
objectTypeName    : IPSet
vsmUuid           : 564D2852-5BAB-0218-392E-B1050109BD46
nodeId            : 47ce48b9-a449-4245-a2a4-
f529c5b83b12
revision          : 1
type              : type
```

```
name                 : IPS-RFC1918
description          : RFC1918 subnets
scope                : scope
clientHandle         :
extendedAttributes :
isUniversal          : false
universalRevision  : 0
inheritanceAllowed : false
value                : 10.0.0.0/8,172.16.0.0/12
```

The newly created IPSet, shown in Example 9.6, can now be used in a new DFW rule, or as a member object in a security group.

NOTE
Additional parameters can be used when creating objects. The `-universal` parameter marks the object available for Universal Replication. The `-scopeId` parameter allows input of an NSX Edge ID to allow use on the NSX Edge firewall.

There will be times where it is required to modify the values (i.e., members) of an existing IP Set. The newly created IP set, shown in Example 9.6, is missing a subnet as per RFC1918. PowerNSX provides the ability to add a new IP address, IP address range or subnet to an existing IP set. Example 9.7 shows the use of `Add-NsxIpSetMember`.

Example 9.7 Appending New IPSet Members

```
PS /> Get-NsxIpSet -name IPS-RFC1918 | Add-
NsxIpSetMember -IPAddress "192.168.0.0/16"

objectId             : ipset-2
objectTypeName       : IPSet
vsmUuid              : 564D2852-5BAB-0218-392E-B1050109BD46
nodeId               : 47ce48b9-a449-4245-a2a4-
f529c5b83b12
revision             : 2
type                 : type
name                 : IPS-RFC1918
description          : RFC1918 subnets
scope                : scope
clientHandle         :
```

```
extendedAttributes :
isUniversal        : false
universalRevision  : 0
inheritanceAllowed : false
value              : 192.168.0.0/16,172.16.0.0/12,10.0.0.0/8
```

Adding `192.168.0.0./16` to the IP set `IPS-RFC1918` means the correct private address space is defined. The IP set is ready for later use in a DFW rule.

Creating Security Groups and Defining Members

One major facet of VMware NSX for vSphere are the objects that can be used as a part of a DFW rule. These objects allow administrators to build scalable abstractions into security that help take the rigidity out of firewall operations.

One such object is the security group. A security group is a construct that is akin to a container. It can have numerous member types within it. Example 9.8 outlines the ease of creating a security group

Example 9.8 Creating a New Security Group

```
PS /> New-NsxSecurityGroup SG-DMZ-Web

objectId           : securitygroup-10
objectTypeName     : SecurityGroup
vsmUuid            : 564D2852-5BAB-0218-392E-B1050109BD46
nodeId             : 47ce48b9-a449-4245-a2a4-
f529c5b83b12
revision           : 1
type               : type
name               : SG-DMZ-Web
description        :
scope              : scope
clientHandle       :
extendedAttributes :
isUniversal        : false
universalRevision  : 0
inheritanceAllowed : false
```

The newly created security group in Example 9.8 is ready to be used. This can be used by a DFW rule and is also an object that can include vCenter objects as child members. The security group itself can also be a member of another security group.

Example 9.9 builds a new security group and retrieves a vCenter cluster object to use for membership.

Example 9.9 Creating a New Security Group Including a Cluster Member

```
PS /> New-NsxSecurityGroup SG-DMZ-Cluster
-IncludeMember (Get-Cluster DMZ-Cluster)

objectId             : securitygroup-11
objectTypeName       : SecurityGroup
vsmUuid              : 564D2852-5BAB-0218-392E-B1050109BD46
nodeId               : 47ce48b9-a449-4245-a2a4-
f529c5b83b12
revision             : 2
type                 : type
name                 : SG-DMZ-Cluster
description          :
scope                : scope
clientHandle         :
extendedAttributes   :
isUniversal          : false
universalRevision    : 0
inheritanceAllowed   : false
member               : member
```

The defined cluster named DMZ-Cluster is now added as a member. This member will include all virtual machines associated with the cluster DMZ-Cluster. An administrator can validate the members of a security group as shown in Example 9.10.

Example 9.10 Discovering Security Group Membership

```
PS /> Get-NsxSecurityGroup SG-DMZ-Cluster | Get-
NsxSecurityGroupEffectiveMember

VirtualMachine IpAddress MacAddress Vnic
-------------- --------- ---------- ----
{vmnodes}                {macNodes} {vnicNodes}
```

The cmdlet `Get-NsxSecurityGroupEffectiveMember` takes a security group object from the pipeline and will validate the VMs that are resolved. In Example 9.10 the VMs that result on cluster DMZ-Cluster are included as properties.

The output of Example 9.10 returns the parent object of the membership. Whilst it is possible to use dot notation to traverse the object and find more detail there are additional cmdlets that achieve the same result.

Example 9.11 demonstrates the output of the various security group membership cmdlets.

Example 9.11 Discovering Security Group Membership Redux

```
PS /> Get-NsxSecurityGroup -name SG-DMZ-Cluster | Get-
NsxSecurityGroupEffectiveVirtualMachine

VmName                                VmId
------                                ----
{win-01, other-01, lnx-02, win-02...} {vm-44, vm-46,
vm-48, vm-45..}

PS /> Get-NsxSecurityGroup -name RFC-TEST | Get-NsxSecu
rityGroupEffectiveIpAddress

IpAddress
---------
{10.0.0.0/8, 192.168.0.0/16, 172.16.0.0/20}

PS /> Get-NsxSecurityGroup -name SG-DMZ-Cluster | Get-
NsxSecurityGroupEffectiveVnic

Uuid
----
{500c5441-eb92-f52b-af4a-adc9f3b3f6d0.000, 500c4eb6-658a-
a24f-f42a-4537fe76bcf3.000, 500c4ff9-8397-9e3f-fb67-
78868fdc646c.000, 500cd...

PS /> Get-NsxSecurityGroup -name SG-DMZ-Cluster | Get-
NsxSecurityGroupEffectiveMacAddress

MacAddress
----------
{00:50:56:8c:58:dc, 00:50:56:8c:83:02, 00:50:56:8c:25:4a,
00:50:56:8c:b3:b3...}
```

Very quickly it becomes clear that there is a lot of detail stored in a membership object. This can be rapidly retrieved thanks to PowerNSX.

In production and environments rules are already created and enforcing security policy. At times rules may need to be edited and in turn objects that provide membership will need updating. Example 9.12 highlights adding a member object to a security group.

Example 9.12 Add Security Group Member

```
PS /> Get-NsxSecurityGroup -name SG-DMZ-Web | Add-
NsxSecurityGroupMember -Member $WebTag

PS /> (Get-NsxSecurityGroup -name SG-DMZ-Web).member.
name
ST-DMZ-Web
```

Example 9.12 above adds the security tag in `$WebTag` variable as a member of the group. When performing a validation against the members of security group SG-DMZ-Web the member property name reveals ST-DMZ-Web.

NOTE
The parameter `–MemberIsExcluded` provides further control over group membership. The parameter –`MemberIsExcluded` provides the ability to exclude an object being if it is a member of another included abstraction. For example, the VMs `VM1, VM2`, and `VM3` reside on cluster `Cluster-A`. Cluster `Cluster-A` is added as a member of security group `SG-DMZ-Cluster`. Virtual machine VM2 should not be a member of security group `SG-DMZ- Cluster`. This is where `–MemberIsExcluded` should be used and as a result it is excluded from the security group.

Just as important as adding members to security groups is the ability to remove. This operations is part of life cycling applications and workloads. Example 9.13 shows the ability to remove an object from a security group.

Example 9.13 Removing a Security Group Member

```
PS /> Get-NsxSecurityGroup SG-DMZ-Web | Remove-
NsxSecurityGroupMember -Member $webtag
```

Example 9.13 removed the security tag `ST-Web` from security group `SG-DMZ-Web`. This can be performed for the following member types listed in Example 9.14

Example 9.14 Security Group Member Types

```
PS /> Get-NsxSecurityGroupMemberTypes
IPSet
ClusterComputeResource
VirtualWire
VirtualMachine
SecurityGroup
DirectoryGroup
VirtualApp
ResourcePool
DistributedVirtualPortgroup
Network
Datacenter
Vnic
SecurityTag
MACSet
```

These types are what are supported as members in security groups.

Working with Security Tags

Security tags allow a string based object to be appended to a VM. These security tags can also be a member criteria for security groups. VMs tagged with a security tag can automatically added to a given group. This results in VMs inheriting new DFW rules based upon their security group membership.

Security tag operations performed with PowerNSX are straight forward. Example 9.15 highlights the use of the `New-NsxSecurityTag` cmdlet.

Example 9.15 Creating New Security Tags

```
PS /> New-NsxSecurityTag -Name ST-Web

objectId            : securitytag-15
objectTypeName      : SecurityTag
vsmUuid             : 564D2852-5BAB-0218-392E-B1050109BD46
nodeId              : 47ce48b9-a449-4245-a2a4-
f529c5b83b12
revision            : 0
```

```
type                : type
name                : ST-Web
clientHandle        :
extendedAttributes :
isUniversal         : false
universalRevision  : 0
systemResource      : false
vmCount             : 0
```

With a new security tag called ST-Web it is now possible to use it as a membership criteria for a security group and apply it to a VM. This step is outlined in Example 9.16.

Example 9.16 Adding VMs to a Security Tag

```
PS /> Get-NsxSecurityTag ST-Web | New-
NsxSecurityTagAssignment -ApplytoVM -VirtualMachine
(get-VM lnx-01)
```

With the security tag ST-Web applied to the VM lnx-01 it will become a member of any security group that uses security tags as a membership criteria. As environments evolve and workloads use various tags a need for security tag usage is apparent. Example 9.17 outlines how to evaluate security tag usage.

Example 9.17 Querying Security Tag Members

```
PS /> Get-NsxSecurityTag ST-Web | Get-
NsxSecurityTagAssignment

SecurityTag VirtualMachine
----------- --------------
securityTag lnx-01
```

The cmdlet in Example 9.17 shows how to determine the VMs associated with a specific security tag.

Removing Security Tags

There comes a time when security lifecycles require an object to be destroyed. The removal of a security tag can have a larger impact as it may be referenced by numerous objects. Example 9.18 highlights an important warning message.

Example 9.18 Removing Security Tags

```
PS /> Get-NsxSecurityTag ST-Web | Remove-NsxSecurityTag

Removal of Security Tags may impact desired Security
Posture and expose your infrastructure. Please
understand the impact of this
change
Proceed with removal of Security Tag ST-Web?
[Y] Yes  [N] No  [?] Help (default is "N"): Y
```

If a security tag is in use as a membership object then removing the security tag will strip VM members of their association to a security group. Furthermore, if that security group is used in a DFW rule or as load balancer pool the policy or membership is revoked. This will lead into a change in security posture and loss of connectivity. Using `Get-NsxSecurityTagAssignment` as demonstrated in Example 9.18 will help ensure the safe removal of security tags.

Creating DFW Rules using Objects

The ability to create a plethora of objects for use in DFW rules with PowerNSX is straightforward. Using them in DFW rules helps simplify security rules in many cases.

NOTE
The upcoming examples use variables for source, destination, and applied to fields for reader ease. Prior examples have shown how to create, retrieve, and store objects in variables for use in cmdlets.

Example 9.19 demonstrates how to provide a mixture of objects for source, destination, and applied to fields.

Example 9.19 Creating Object-Based DFW Rules

```
Get-NsxFirewallSection DMZ | New-NsxFirewallRule -name
"Internet Access" -source $LS -destination $ipsrfc1918
-NegateDestination   -Action "allow" -AppliedTo
$sgcluster -Service $httptraffic

id            : 1010
disabled      : false
logged        : false
name          : Internet Access
action        : allow
appliedToList : appliedToList
sectionId     : 1005
sources       : sources
destinations  : destinations
services      : services
direction     : inout
packetType    : any
```

This firewall rule uses three different object types. They are a logical switch, IP set, and a security group. This firewall rule is designed to allow VMs on the logical switch defined by the parameter `-Source` to the IP set defined by the `-Destination` parameter. This IP set uses a `-NegateDestination` and as a result it will be all destinations except those defined in the IP set. The parameter `-Action` will allow traffic and the defined `-Service` parameter is for HTTP/S traffic. The DFW rule is applied the vNIC of all VMs in the security group defined in the `-AppliedTo` property.

NOTE

There are hundreds, if not thousands, of permeations of rules and examples that could be expressed here. The goal of this section is to give an insight into what can be used and how it could be used. Please utilize `Get-Help | New-NsxFirewallRule -examples` for examples or visit https://powernsx.github.io for more details.

To validate this rule is applied correctly to an expected VM it is possible to use a PowerNSX cmdlet called `Get-NsxCliDfwRule`. This allows a VM to be defined in a parameter and results in retrieving data plane configuration. Example 9.20 highlights this.

Example 9.20 Validate DFW Rule on vNIC

```
PS /> Get-NsxCliDfwRule -VirtualMachine $lnx02
WARNING: This cmdlet is experimental and has not been
well tested.  Its use should be limited to
troubleshooting purposes only.

RuleSet         : domain-c7
InternalRule    : False
RuleID          : 1010
Position        : 1
Direction       : inout
Type            : Layer3
Service         : tcp
Source          : addrset ip-virtualwire-7
Destination     : not addrset ip-ipset-2
Port            : 80
Action          : accept
Log             : False
Tag             :
VirtualMachine  : lnx-02
Filter          : nic-123791-eth0-vmware-sfw.2
```

The rule has been successfully pushed to the filter of VM `lnx-02`. It matches the recently created rule in Example 9.19.

Adding a new source or destination object can be required as applications or DFW rules change. Whilst it is possible to edit a security group membership there are occasions where an entire new object is required. Example 9.21 outlines how to add a new object to an existing DFW rule.

Example 9.21 Add a New DFW Rule Object

```
PS /> Get-NsxFirewallSection DMZ | Get-NsxFirewallRule
"Internet Access" | Add-NsxFirewallRuleMember
-MemberType Source -Member $DMZViewLogicalSwitch

RuleId     : 1011
SectionId  : 1005
MemberType : Source
Name       : LS-DMZ
Value      : virtualwire-7
Type       : VirtualWire
isValid    : true
```

```
RuleId      : 1011
SectionId   : 1005
MemberType  : Source
Name        : DMZ-View-Connection-Servers
Value       : virtualwire-8
Type        : VirtualWire
isValid     : true

RuleId      : 1011
SectionId   : 1005
MemberType  : Destination
Name        : IPS-RFC1918
Value       : ipset-2
Type        : IPSet
isValid     : true
```

The output of Example 9.21 shows a new logical switch `DMZ-View-Connection-Servers` as a source added for the DFW rule `1011`. The existing objects are preserved and printed to the screen as validation.

NOTE

The parameter `–MemberType` allows definition of `Destination`, `Source`, or `Both`. This allows the object being appended to be added to either fields or both in one action.

Removing an object from a DFW can be performed with a similar cmdlet. Example 9.22 removes an old logical switch from a DFW rule.

Example 9.22 Remove a DFW Rule Object

```
PS /> Get-NsxFirewallSection DMZ | Get-NsxFirewallRule
"Internet Access" | Get-NsxFirewallRuleMember -Member
LS-DMZ | Remove-NsxFirewallRuleMember
Removal of a firewall rule member is permanent and
will modify your security posture.
Proceed with removal of member virtualwire-7 from the
Source list of firewallrule 1011 in section 1005?
[Y] Yes  [N] No  [?] Help (default is "N"): Y

PS /> Get-NsxFirewallSection DMZ | Get-NsxFirewallRule
"Internet Access" | Get-NsxFirewallRuleMember -Member
LS-DMZ
```

The output in Example 9.22 will pipe the retrieved DFW rule member to `Remove-NsxFirewallRuleMember`. This results in the member being removed from the firewall rule. Using `Get-NsxFirewallRuleMember` against the specific rule `Internet Access` validates the member `LS-DMZ` is now removed.

Progressive Example: Creating Security Objects and DFW Rules

The progressive example provides additional examples related to a specific topology. It is built upon in each successive chapter within this book.

The diagram in Figure 9.1 highlight the logical security groups applied to the example topology.

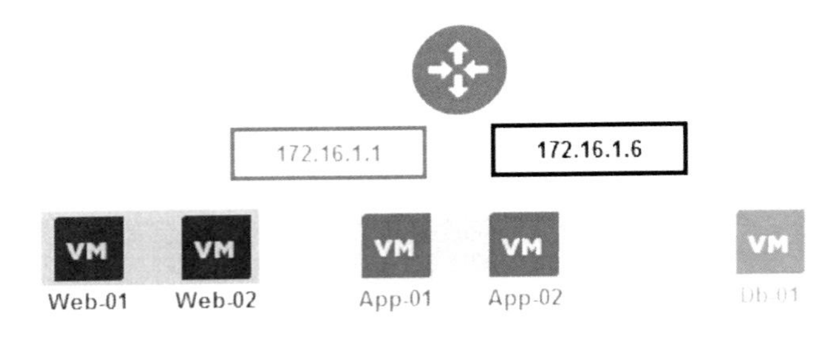

Figure 9.1 Logical Security Topology

Below are the traffic flows that must be secured:

1. Source NAT from ESG vNIC on transit network to web tier on `TCP 80`
2. Web tier to app load balancer IP address on `TCP 80`
3. Source NAT from ESG vNIC on transit network to app tier on `TCP 80`
4. App tier to DB tier on `TCP 3306`

With the topology defined it is time to create the objects that are required. Example 9.23 builds the required objects for this example.

Example 9.23 Logical Security Objects

```
PS /> $IpsEsgPri = New-NsxIpset -Name "ESG-TRANSIT-PRI"
-IPAddresses 172.16.1.1
PS /> $IpsEsgSec = New-NsxIpset -Name "ESG-TRANSIT-SEC"
-IPAddresses 172.16.1.6
PS /> $WebSg = New-NsxSecurityGroup -name "SG-Web"
-IncludeMember $Web01,$Web02
PS /> $AppSg = New-NsxSecurityGroup -name "SG-App"
-IncludeMember $App01,$App02
PS /> $DbSg = New-NsxSecurityGroup -name "SG-Db"
-IncludeMember $Db01
```

Example 9.23 outlines the security groups and the IP sets that need to be created for the DFW rules. Example 9.24 demonstrates using the objects created to create a DFW rule between the ESG and the web tier VMs.

Example 9.24 Load Balancer to Web Tier

```
PS /> Get-NsxFirewallSection $Section.name | New-
NsxFirewallRule -name "LB to Web Tier" -Source
$IpsEsgPri -Destination $WebSg -Service $http -Action
allow -appliedTo $WebSg -EnableLogging -Tag "LbToWeb"
-Comment "LB to Web Tier"

id             : 1139
disabled       : false
logged         : true
name           : LB to Web Tier
action         : allow
notes          : LB to Web Tier
appliedToList  : appliedToList
sectionId      : 1112
sources        : sources
destinations   : destinations
services       : services
direction      : inout
packetType     : any
tag            : LbToWeb
```

Example 9.25 demonstrates using the objects created to create a DFW rule between the ESG and the web tier VMs.

Example 9.25 Web to App VIP

```
PS /> Get-NsxFirewallSection $Section.name | New-
NsxFirewallRule -name "Web Tier to App LB VIP" -Source
$WebSg -Destination $IpsEsgSec -Service $http -Action
allow -AppliedTo $WebSg -EnableLogging -Tag
"WebToAppVIP" -Comment "Web Tier to App LB VIP"

id              : 1140
disabled        : false
logged          : true
name            : Web Tier to App LB VIP
action          : allow
notes           : Web Tier to App LB VIP
appliedToList   : appliedToList
sectionId       : 1112
sources         : sources
destinations    : destinations
services        : services
direction       : inout
packetType      : any
tag             : WebToAppVIP
```

Example 9.25 outputs the rule created for access from the Web Tier to the App VIP on the ESG. Example 9.26 highlights access from the App VIP to the App tier. Due to source NAT being used on the ESG's VIP, the IP address that sources the traffic is the vNIC connected to the network. As a result, the primary IP address on interface is the source of the traffic. In this case, it is `172.16.1.1` and not the VIP `172.16.1.6`.

Example 9.26 App VIP / ESG Interface to App Tier

```
PS /> Get-NsxFirewallSection $Section.name | New-
NsxFirewallRule -name "App VIP to App Tier" -Source
$IpsEsgPri -Destination $AppSg -Service $http -Action
allow -AppliedTo $AppSG -EnableLogging -Tag
"AppVIPToApp" -Comment "App VIP to App Tier"
```

```
id             : 1141
disabled       : false
logged         : true
name           : App VIP to App Tier
action         : allow
notes          : App VIP to App Tier
appliedToList  : appliedToList
sectionId      : 1112
sources        : sources
destinations   : destinations
services       : services
direction      : inout
packetType     : any
tag            : AppVIPToApp
```

The rule created in Example 9.26 allows HTTP traffic from the defined source to the app tier. Example 9.27 opens the last key flow to the database tier.

Example 9.27 App Tier to DB Tier

```
PS /> Get-NsxFirewallSection $Section.name | New-
NsxFirewallRule -Name "App Tier to DB Tier" -Source
$AppSg -Destination $DbSg -Service $mysql -Action
allow -AppliedTo $AppSG,$DbSg -EnableLogging -Tag
"AppToDb" -Comment "App Tier to DB Tier"

id             : 1142
disabled       : false
logged         : true
name           : App Tier to DB Tier
action         : allow
notes          : App Tier to DB Tier
appliedToList  : appliedToList
sectionId      : 1112
sources        : sources
destinations   : destinations
services       : services
direction      : inout
packetType     : any
tag            : AppToDb
```

With the explicitly allowed traffic for the application allowed through the application needs to be secured. This is done in Example 9.28 with a deny rule that is applied to all three tiers.

Example 9.28 Application Specific Deny

```
PS /> Get-NsxFirewallSection $Section.name | New-
NsxFirewallRule -Name "Progressive Example Deny"
-Service $TcpAll,$UdpAll -Action deny -AppliedTo
$WebSg,$AppSg,$DbSg -EnableLogging -Tag "3taDrop"
-Comment "Deny rule for Progressive Example" -Position
Bottom

id             : 1143
disabled       : false
logged         : true
name           : Progressive Example Deny
action         : deny
notes          : Deny rule for Progressive Example
appliedToList  : appliedToList
sectionId      : 1112
services       : services
direction      : inout
packetType     : any
tag            : 3taDrop
```

Example 9.28 applies the final rule to the progressive example. It will block all TCP and UDP traffic to all application tiers from unspecified resources. This results in a secure application topology per the requirements that has been deployed and configured by PowerNSX.

CAUTION

Creating new DFW rules with PowerNSX will place it at the top of the referenced section. When creating a new deny rule this could result legitimate traffic being dropped. Use the `-Position` parameter to place the rule at the bottom of a given section. The parameter `-Position` takes the input of `Top` or `Bottom`.

Chapter Summary

This chapter has introduced tasks and operations around DFW. An administrator using PowerNSX should now be confident in creating, updating, and removing DFW rules, NSX security objects, and firewall sections.

These skills will serve as a foundation to managing the security posture of an environment with PowerNSX.

Chapter 10

Cross vCenter and PowerNSX

VMware NSX for vSphere can be configured in a Cross vCenter configuration. This allows constructs known as universal objects to be created. These universal objects are replicated between NSX Managers participating in the cross vCenter NSX environment.

A substantial update was made to PowerNSX so it could be used in Cross vCenter environments. PowerNSX supports the ability to create, modify, and delete universal objects.

Design Considerations

The design consideration around handling universal objects was key before the functionality was created. Early on, an idea to create separate cmdlets was discussed and raised. It was quickly decided that this would lead to a substantial amount of duplication and effort. The result was to utilize the existing cmdlets and update them to work with cross vCenter NSX. The cmdlets that have a universal option within VMware NSX for vSphere now have an additional `-Universal` or `-IsUniversal` parameter appended.

This allows a consistent approach in naming convention and usage when using local or universal objects with the PowerNSX cmdlets.

Retrieving Universal Objects

The implementation of cross vCenter functionality into PowerNSX did impact the standard behavior of some cmdlets. Example 10.1 highlights retrieving transport zones.

Example 10.1 Retrieving Transport Zones

```
PS /> Get-NsxTransportZone | Select-Object Name, id

name id
---- --
UTZ1 universalvdnscope
TZ1  vdnscope-1
```

All transport zones are returned when using `Get-NsxTransportZone`. This returns a local and universal transport.

The use of `-LocalOnly` or `-UniversalOnly` will help filter the results to ensure the correct objects are retrieved. Example 10.2 demonstrates this.

Example 10.2 Retrieving Universal Transport Zone

```
PS /> Get-NsxTransportZone -UniversalOnly | Select-
Object Name, id

name id
---- --
UTZ1 universalvdnscope
```

With a universal transport zone retrieved it can now be used. This can be piped to the `New-NsxLogicalSwitch` cmdlet to when creating a new universal logical switch as shown in Example 10.3.

Example 10.3 Creating Universal Logical Switch

```
PS /> Get-NsxTransportZone -UniversalOnly | New-
NsxLogicalSwitch ULS-App1


objectId                : universalwire-14
objectTypeName          : VirtualWire
vsmUuid                 : 4201B045-B1F9-457F-E621-
B54038A6AFA5
nodeId                  : 4b749a6a-bc41-431b-bf24-
cf9e54dcb452
revision                : 3
type                    : type
name                    : ULS-App1
description             :
clientHandle            :
extendedAttributes      :
isUniversal             : true
universalRevision       : 3
tenantId                :
vdnScopeId              : universalvdnscope
vdsContextWithBacking : {vdsContextWithBacking,
vdsContextWithBacking}
vdnId                   : 10012
guestVlanAllowed        : false
controlPlaneMode        : UNICAST_MODE
ctrlLsUuid              : 782865e9-441b-4667-9454-
b02685719cde
macLearningEnabled      : false
```

The universal transport zone was piped to `New-NsxLogicalSwitch` and as a result a new universal logical switch called `ULS-App1` was made.

NOTE
An observant reader will note that there is not `-Universal` parameter used on `New-NsxLogicalSwitch`. This is due to the transport zone type denoting the type of logical switch. Local transport zones allow logical switches to be created whilst universal transport zones allow universal logical switches to be created.

Creating Universal Objects

Creating universal objects with PowerNSX is like creating local objects with PowerNSX. Functions that support cross vCenter NSX have an additional parameter to indicate if it is to be made universal. Example 10.4 demonstrates creating a universal logical router.

Example 10.4 Creating Universal Logical Router

```
PS /> New-NsxLogicalRouter -name DLR-Universal -Tenant
coke -ManagementPortGroup $pg -Cluster $cl -Datastore
$ds -Interface $lif1, $lif2, $lif3, $uplif -Universal



id                   : edge-b7962130-3cc0-43aa-a02a-
ffa32cb10968
version              : 2
status               : deployed
tenant               : coke
name                 : DLR-Universal
fqdn                 : NSX-edge-b7962130-3cc0-43aa-a02a-
ffa32cb10968
enableAesni          : true
enableFips           : false
vseLogLevel          : info
appliances           : appliances
cliSettings          : cliSettings
features             : features
autoConfiguration    : autoConfiguration
type                 : distributedRouter
isUniversal          : true
universalVersion     : 0
mgmtInterface        : mgmtInterface
interfaces           : interfaces
edgeAssistId         : 10000
lrouterUuid          : d1639657-4537-498d-ba03-
98b5ebf1dc27
queryDaemon          : queryDaemon
localEgressEnabled   : false
edgeSummary          : edgeSummary
```

The only difference in the `New-NsxLogicalrouter` cmdlet to create universal logical router versus a logical router is the `-Universal` parameter.

NOTE

The interface specs used in Example 10.4 have universal logical switches associated to them.

When creating a universal IP set the same `-Universal` parameter is used.

Example 10.5 Creating Universal IP Set

```
PS /> New-NsxIpSet -name UIPS-RFC1918 -IPAddresses "10.0
.0.0/8,192.168.0.0/16,172.16.0.0/12" -Universal


objectId             : ipset-84999f44-3dd1-49bd-984c-
0d2eab95b938
objectTypeName       : IPSet
vsmUuid              : 4201B045-B1F9-457F-E621-
B54038A6AFA5
nodeId               : 4b749a6a-bc41-431b-bf24-
cf9e54dcb452
revision             : 1
type                 : type
name                 : UIPS-RFC1918
description          :
scope                : scope
clientHandle         :
extendedAttributes   :
isUniversal          : true
universalRevision    : 0
inheritanceAllowed   : false
value                : 172.16.0.0/12,10.0.0.0/8,192.168.0.0/16
```

The newly created Universal IPSet has the value of `true` in the `isUniversal` property.

Chapter Summary

This chapter has introduced some considerations and design decisions the PowerNSX team made when dealing with universal objects and cross vCenter NSX. Understanding these decisions and considerations will aid in using PowerNSX with universal objects. An administrator using PowerNSX should now be confident in administering and operating cross vCenter NSX environments.

Chapter 11

Administrative Operations

PowerNSX provides administrators an additional tool to aid with operational tasks. The following chapter outlines how PowerNSX can help with common operational tasks performed by an NSX administrator. These are in addition to the create, retrieve, update, delete (CRUD) operations demonstrated in previous chapters.

Searching for a Port

Searching for the correct service can be finding a needle in a haystack. This can be efficiently achieved with the `Get-NsxService` cmdlet. The cmdlet allows searching through using the `-port` parameter a feature than until very recently was not possible in the NSX UI.

Example 11.1 demonstrates the search for port `80`.

Example 11.1 Find Service by Port

```
name                                                            isUniversal
----                                                            -----------

Horizon 6 Connection Server to View Composer Service communication   false
Horizon 6 Default HTTPS Client to Connection and Security Servers    false
Horizon 6 Connection Server to vCenter Server communication          false
HTTP                                                                 false
HTTP                                                                 true
Horizon 6 Connection Server to View Composer Service communication   true
Horizon 6 Default HTTPS Client to Connection and Security Servers    true
Horizon 6 Connection Server to vCenter Server communication          true
```

This command returns all ports that specifically have a port match. This will return both `Universal` and `Local` services. The `Select-Object` statement filters the `name` and `isUniversal` properties as they are most pertinent.

Services are matched, even if the port number falls within a port range, not just an explicit port number. Example 11.2 finds port `8032` within a port range.

Example 11.2 Find Service by Port within a Range

```
PS /> Get-NsxService -port 8032 | Select-Object name, isUniversal

name                                                                isUniversal
----                                                                ----
Win - RPC, DCOM, EPM, DRSUAPI, NetLogonR, SamR, FRS - UDP          false
Win - RPC, DCOM, EPM, DRSUAPI, NetLogonR, SamR, FRS - TCP          false
VMware-VDM2.x-Ephemeral                                             false
Win - RPC, DCOM, EPM, DRSUAPI, NetLogonR, SamR, FRS - UDP          true
Win - RPC, DCOM, EPM, DRSUAPI, NetLogonR, SamR, FRS - TCP          true
VMware-VDM2.x-Ephemeral                                             true

PS /> (Get- NsxService -name "VMware-VDM2.x-Ephemeral").element

applicationProtocol value
------------------- -----
TCP                 1024-65535
```

By searching for port `8032` it returns six services that contain the port `8032`. Expanding the element property of VMware-VDM2.x-Ephemeral shows the service uses a range in the value property. The port 8032 lies within the range of 1024-65535.

Does a Firewall Rule Encompass a Specific Address?

There are times when an administrator will need to determine if a specific address or set of addresses are covered by a firewall rule. This can occur when provisioning a new workload or troubleshooting connectivity. Example 11-3 demonstrates finding a specific address in the destination parameter of `Get-NsxFirewallRule`.

Example 11.3 Find Address used in a DFW Rule

```
PS /> Get-NsxFirewallRule -Destination "192.168.103.100"
| select name

name                     id
----                     --
Progressive Example Deny 1010
Default Rule NDP         1003
Default Rule DHCP        1002
Default Rule             1001
```

The IP address has been detected in the destination field of four rules. The firewall rule `Internet Access` is an IP set that has three subnets defined from RFC1918.

The returned rules based on the `-Destination "192.168.103.100"` are found within explicitly defined IP addresses, IP ranges, or VM objects. These are sourced from learned the translation of learned IP addresses to VM objects. Example 11.4 demonstrates this.

Example 11.4 Find VM used in DFW Rule

```
PS /> Get-NsxFirewallRule -source $vm | select name,
id

name                         id
----                         --
Windows Workload             1011
Progressive Example Deny 1010
Default Rule NDP             1003
Default Rule DHCP            1002
Default Rule                 1001
```

The `Get-NsxFirewallRule` cmdlet is passed a VM within a variable on the `-source` parameter. The VM is revealed to be a member of five DFW rules.

NOTE
It is also possible to use the parameter `-source` or `-both` alongside `-destination` to further control or scope this lookup. It is important to note that this query leverages the NSX Manager to do the heavy lifting of the translation between IP or VM, and the source or destination of a rule that causes the hit. Rules will be hit, even if they list a security group that indirectly has a VM object as a member when specifying the IP (that NSX has learned) used by that VM.

Environments change and as such firewall rules must adapt. There are cmdlets designed for operations pertaining to adding and removing DFW rule members. Example 11.5 demonstrates adding a new destination member.

Example 11.5 Find Address used in a DFW Rule

```
PS /> Get-NsxFirewallSection $section.name | Get-
NsxFirewallRule -name "DMZ Rule" | Add-
NsxFirewallRuleMember -MemberType Destination -Member
$NewDMZSecurityGroup

RuleId     : 1145
SectionId  : 1113
MemberType : Source
Name       : IPS-RFC1918
Value      : ipset-2
Type       : IPSet
isValid    : true

RuleId     : 1145
SectionId  : 1113
MemberType : Destination
Name       : DMZ-Rule
Value      : securitygroup-18
Type       : SecurityGroup
isValid    : true

RuleId     : 1145
SectionId  : 1113
MemberType : Destination
Name       : New-DMZ-WebApp
Value      : securitygroup-19
Type       : SecurityGroup
isValid    : true
```

The newly added member was a destination. The security group New-DMZ-WebApp was added to the DFW rule "DMZ Rule".

Note

The `-MemberType` property can be `Source`, `Destination` or `Both`.

Throughout firewall operations it is possible to inadvertently remove the remaining member of a DFW rule. Example 11.6 demonstrates what occurs these scenarios.

Example 11.6 Removing a Rule Source or Destination Member

```
PS /> Get-NsxFirewallSection $section.name | Get-
NsxFirewallRule -name "DMZ Rule" | Get-
NsxFirewallRuleMember -member DMZ-Rule | Remove-
NsxFirewallRuleMember

Removal of a firewall rule member is permanent and
will modify your security posture.

Proceed with removal of member securitygroup-18 from
the Destination list of firewallrule 1145 in section
1113?
[Y] Yes  [N] No  [?] Help (default is "N"): Y

The destination member securitygroup-18 of rule 1145
in section 1113 is the last destination member in this
rule.  Its removal will cause this rule to match ANY
Destination

Confirm rule 1145 to match Destination ANY?
[Y] Yes  [N] No  [?] Help (default is "N"): Y

WARNING: The destination member securitygroup-18 of
rule 1145 in section 1113 was the last member in this
rule.  Its removal has caused this rule to now match
ANY Destination.
```

Example 11.6 demonstrates the removal of a member from a DFW rule. This is the last object in the destination list and it will throw a warning to the administrator. Extra confirmation is required to ensure that the administrator acknowledges that the rule will match any destination - effectively meaning it applies to everything, rather than just one thing that it did previously.

This is an example of how PowerNSX provides additional smarts and safeguards to firewall operations.

Cloning an Existing NSX Edge

The cmdlet `Copy-NsxEdge` provides administrators the ability to quickly replicate an existing Edge and its configuration. This lends itself to scenarios such as validation, configuration testing, or scale out. Example 11.7 demonstrates how the `Copy-NsxEdge` cmdlet operates.

Example 11.7 Copying an Existing NSX Edge

```
PS /> Get-NsxEdge ecmp-edge1 | Copy-NsxEdge
Supply values for the following parameters:
Name: ecmp-edge-3
Password: VMware1!VMware1!

WARNING: IPSec PSK for site global set to BZ1A0icr.
Please update manually as required.

Enter new primary address for source edge addressgroup
with existing IP 172.16.10.11 on vnic 0: 172.16.10.17
Enter new primary address for source edge addressgroup
with existing IP 172.16.20.11 on vnic 1: 172.16.20.17

WARNING: Updating Router ID. Previous ID : 172.16.10.11,
Updated ID : 172.16.10.17
WARNING: Performing firewall fixups for any user based
rules that contained local object references on
edge-40.

id : edge-40
version : 2
status : deployed
tenant : default
name : ecmp-edge3
fqdn : ecmp-edge3
enableAesni : true
enableFips : false
vseLogLevel : info
vnics : vnics
appliances : appliances
cliSettings : cliSettings
features : features
autoConfiguration : autoConfiguration
type : gatewayServices
isUniversal : false
hypervisorAssist : false
queryDaemon : queryDaemon
edgeSummary : edgeSummary
```

The cmdlet will validate the configuration of the existing NSX Edge passed along the pipeline from `Get-NsxEdge`. It assesses properties such as Interface IPs, NSX Edge name, and other configuration on the edge. Where required the cmdlet will prompt the administrator to resolve it.

`Copy-NsxEdge` actually does a lot of heavy lifting and fixups to properly and completely duplicate an Edge configuration. This includes regenerating self-signed certificates, reconfiguring all listening services (e.g., load balancing, SSL VPN, DHCP) to bind to the new NSX Edge's IP, reconfiguring router IDs, recreating locally scoped objects like IP sets services, and fixing NAT or NSX Edge firewall rules that reference any NSX Edge interface address.

NOTE
An example use of this cmdlet can be to help scale a 2-node ECMP NSX Edge cluster to 8-nodes quickly. The administrator would specify new interface IP addresses and the rest of the configuration would be copied across.

Searching Firewall Rules for Log Status

Whether an administrator has a few dozen or a few thousand rules it is critical that information can be accessed readily. Example 11.8 highlights how to retrieve rules that are not logging.

Example 11.8 Finding Firewall Rules without Logging Enabled

```
PS /> Get-NsxFirewallSection | Get-NsxFirewallRule |Where-
Object{$ _ .logged -eq "false"} | Select id,name,logged,tag

id   name                         logged  tag
--   ----                         ------  ---
1010 Progressive Example Deny     false
1081 DMZ-App-Stormwind            false   SW-DMZ-APP-allow
1032 DMZ-App-Elune                false   EL-DMZ-APP-allow
1001 Default Rule                 false
```

Across all firewall sections and firewall rules only those that have the property `-logged` and the value of `false` are returned. Using `Select` to output select parameters an administrator can easily see relevant information regarding logging. The addition of the `id`, `name`, and `tag` properties provide relatable information in the output.

Retrieving Firewall Rules with a Specific Tag

Firewall rules may not appear to be related or be grouped in the same section. A DFW tag may be used to group common rules. Example 11.9 retrieves all rules with the same DFW tag.

Example 11.9 Retrieving a Firewall Rule Based on Tag Name

```
PS /> Get-NsxFirewallSection | Get-NsxFirewallRule |
Where-Object {$_.tag -eq "g"}

id             : 1007
disabled       : false
logged         : true
name           : App VIP to App Tier
action         : allow
notes          : App VIP to App Tier
appliedToList  : appliedToList
sectionId      : 1004
sources        : sources
destinations   : destinations
services       : services
direction      : inout
packetType     : any
tag            : AppVIPToApp
```

By using these cmdlets coupled with a `Where-Object` statement it allows the administrator to check rules based on a common property.

Chapter Summary

This chapter has introduced tasks and operations that an NSX operator using PowerNSX may see throughout their day. An administrator using PowerNSX should be confident in leveraging PowerShell to retrieve, filter, create, and manage infrastructure.

These skills will serve not only in managing an VMware NSX for vSphere environment but the infrastructure as well.

Chapter 12

Tools built with PowerNSX

PowerNSX is a superb standalone tool to interact with VMware NSX for vSphere. It can also be used to create additional tools.

NSX Capture Bundle Tool

The NSX bundle capture tool collects data about an NSX installation and stores in for offline use. It has a 'point in time' snapshot of the NSX environment. The tool will interrogate all objects and constructs and store the XML output of them. It currently uses the PowerShell CLIXML format for export information. These are stored in function specific files for traversal and use by other tools and applications.

NOTE

The capture bundle tool is a raw data capture tool. It is currently used in more advanced scenarios as a source of data input for tooling. The bundle capture can also be used as a point of reference to describe an environment if remote-access cannot be granted. The PowerNSX team understand that it is not very approachable in its current state. They plan to build a friendlier interface in the future. Some examples of tools that use the capture bundle are the Visio diagramming tool.

Example 12.1 shows the steps required to run the capture script.

Example 12.1 Object Capture Bundle

```
PowerCLI C:\> .\NsxObjectCapture.ps1
PowerNSX Object Capture Script

Getting NSX Objects
  Getting LogicalSwitches
  Getting DV PortGroups
  Getting VSS PortGroups
  Getting Logical Routers
  Getting Edges
  Getting NSX Controllers
  Getting VMs
  Getting IP and MAC details from Spoofguard

Creating Object Export Bundle

Capture Bundle created at C:\Users\Administrator\
Documents\VMware\NSXObjectCapt
re\NSX-ObjectCapt
ure-192.168.101.201-2017 _ 07 _ 04 _ 03 _ 50 _ 40.zip
```

With the object capture bundle having collected it is possible to import them and use them by other tools. Administrators can manually traverse the content of the bundle for their own purposes. Furthermore, tools can use this data as input for perform other actions.

After unzipping the bundle, it is possible to import a given resource into PowerShell. Example 12.2 demonstrates how.

Example 12.2 Bundle Import

```
PS /> $edge = Import-Clixml $dir/EdgeExport.xml
```

The content of `$dir` is the working direction used for brevity in the example. The file `EdgeExport.xml` is stored in the variable `$edge`. It contains all NSX Edges in the environment including their configuration for all settings and features enabled. This includes but not limited to routing, load balancing, SSL VPN, interfaces, and edge configuration.

Example 12.3 shows how to traverse the contents stored within the variable `$edge` and make it usable.

Example 12.3 Traversing the Bundle

```
PS /> $edge
Name                              Value
----                              -----
edge-7                            <edge><id>edge-7</id>...
edge-3                            <edge><id>edge-3</id...
edge-8                            <edge><id>edge-8</id...

PS /> [xml]$edge7= $edge.Item("edge-7")
PS /> $edge7.edge

id                   : edge-7
version              : 3
status               : deployed
tenant               : default
name                 : Edge1
fqdn                 : Edge1
enableAesni          : true
enableFips           : false
vseLogLevel          : info
vnics                : vnics
appliances           : appliances
cliSettings          : cliSettings
features             : features
```

```
autoConfiguration : autoConfiguration
type              : gatewayServices
isUniversal       : false
hypervisorAssist  : false
queryDaemon       : queryDaemon
edgeSummary       : edgeSummary
```

The first step sees the content of `$edge` list all edges within a construct known as a hash table. With interest in a specific edge the content is `edge-7` is called via item named `edge-7` and stored in a variable `$edge7`. The `[xml]` indicates that the variable `$edge7` is an XML document.

When expanding the edge property of `$edge7` the content of the given edge is contained. It yields the same result querying NSX Manager with `Get-NsxEdge` using the `-objectId` parameter.

NOTE

At the time of writing the object capture bundle collections logical topology details. Keep checking the `vmware/powernsx` repo on GitHub for the latest supported features and functions.

Visio Diagramming Tool

Documentation can be tough. Keeping it up to date is even harder especially given the context of virtualized networking. Using PowerNSX, PowerCLI, and some PowerShell it is possible to automate the documentation of environments using the contents of the capture bundle.

The Visio diagramming tool takes the contents of the capture bundle and creates a diagram of the logical environment. By driving a Visio API and using the capture bundle data, the script can accurately and efficiently build the configuration of the NSX environment in Visio.

Example 12.4 demonstrates how to use a capture bundle with the NSX Visio diagram tool.

Example 12.4 Traversing the Bundle

```
PowerCLI C:\> .\NsxObjectDiagram.ps1 -CaptureBundle
C:\$BundlePath\NSX-ObjectCapt
ure-192.168.100.2017_07_20_15_34_25.zip
PowerNSX Object Diagram Script

Launching Microsoft Visio.

Building Diagram
  Adding nsx-m-01a to diagram with stencil Manager
  Adding Internal to diagram with stencil PortGroup
  Connecting Manager with PortGroup with text:
192.168.100.201
  Adding App02 to diagram with stencil VM Basic
  Adding App to diagram with stencil logical switch
  Connecting VM Basic to logical switch with text:
  Adding App01 to diagram with stencil VM Basic
  Connecting VM Basic.7 with logical switch with text:
  Adding NSX_Controller_350007a3-f3a2-47a6-8455-
b2512591cf7a to diagram with stencil    Controller
  Connecting Controller with PortGroup with text:
192.168.100.202
  Adding DB01 to diagram with stencil VM Basic
  Adding Db to diagram with stencil logical switch
  Connecting VM Basic.11 with logical switch.12 with
text:
  Adding Web01 to diagram with stencil VM Basic
  Adding Web to diagram with stencil logical switch
  Connecting VM Basic.14 with logical switch.15 with
text:
  Adding Web02 to diagram with stencil VM Basic
  Connecting VM Basic.17 with logical switch.15 with
text:
  Adding Dlr01 to diagram with stencil Logical Router
  Adding Transit to diagram with stencil Logical
Router
  Connecting Logical Router with logical switch.20
with text: 172.16.1.2
  Connecting Logical Router with logical switch.15
with text: 10.0.1.1
  Connecting Logical Router with logical switch with
text: 10.0.2.1
  Connecting Logical Router with logical switch.12
with text: 10.0.3.1
```

```
Adding Edge01 to diagram with stencil Edge
  Connecting Edge with PortGroup with text: uplink:
192.168.100.192 192.168.100.193
  Connecting Edge with logical switch.20 with text:
internal: 172.16.1.1 172.16.1.6

Saved diagram at C:\Users\Administrator\Documents\NSX-
ObjectCapture-192.168.100.2017_07_20_15_34_25.vsdx
```

Whilst the tools progress is being output to the console, in the background the Visio diagram is being populated.

NOTE
The API used to drive Visio are quite dated. The authors have found that minimizing the Visio application whilst running this tool results in a substantial performance increase.

Figure 12.1 shows the final diagram built in Visio.

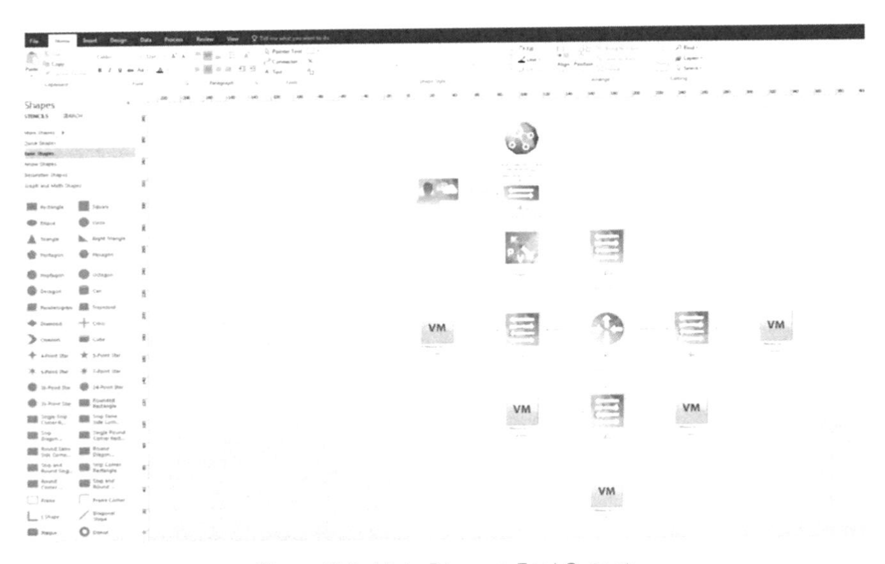

Figure 12.1 Visio Diagram Tool Output

The Visio diagram is built from the capture bundle. This point in time snapshot of the environment configuration ensures details about the environment are accurate. In turn, the automated process of the Visio diagram tool ensures an exact diagram based on the contents of the capture bundle.

DFW to Excel Documentation Tool

Tony Sangha, practice lead for NSX professional services in ANZ, wrote a tool that will create an Excel spreadsheet with detailed DFW configuration. It captures security groups, security group membership, security tags, security group entity types, IP sets, MAC sets, services, service groups, layer 3 firewall rules, and excluded workloads.

An administrator will need to use the Excel spreadsheet provided to ensure the `DFW2Excel.ps1` script can populate the correct fields. The script will perform several API calls using PowerNSX to populate the spreadsheet. Figure 12.2 shows the output of a completed spreadsheet.

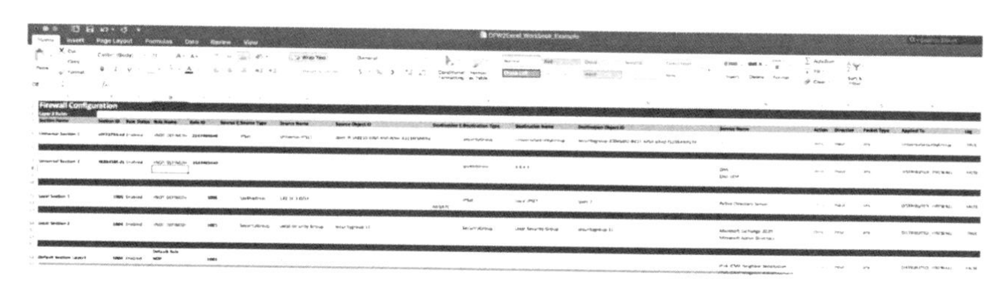

Figure 12.2 Completed PowerNSX

Readers should look at Tony's GitHub repository at https://github.com/tonysangha/PowerNSX-DFW2Excel to get started.

Build NSX from Scratch

Build NSX from scratch has been around from the early days of PowerNSX. It was one of the first true examples of the power of automation. PowerNSX and PowerCLI provide a unified approach to automating vSphere and NSX environments. This will allow an administrator to stand a new NSX environment up within 10 minutes without human interaction.

The script has three methods of operation. Example 12.5 shows the first option.

Example 12.5 Deploy NSX

```
PS /> ./NsxBuildFromScratch.ps1 -deploy3ta:$false
```

Using the `-deploy3ta:$false` switch will execute the script excluding the 3 tier application portion. This results in an NSX environment being built from scratch.

Example 12.6 uses a different switch.

Example 12.6 Deploy only 3 Tier Application

```
PS /> ./NsxBuildFromScratch.ps1 -buildnsx:$false
```

The switch `-buildnsx:$false` will ensure only the 3 tier application is built and deployed. Example 12.7 builds both an NSX environment and deploys the 3 tier application.

Example 12.7 Deploy NSX and 3 Tier Application

```
PS /> ./NsxBuildFromScratch.ps1
```

With the deployment finished the reader can now start using their NSX environment.

An administrator can modify the variables within the script to tailor it to an environment. This can be used to deploy numerous environments if the reader is an integrator or VMware partner. If the reader has a lab environment they can use the `NsxBuildFromScratch` script to ensure consistent deployment for test and validation scenarios.

Readers should look at the official PowerNSX repository at https://github.com/vmware/powernsx/blob/master/Examples/NSXBuildFromScratch.ps1 for the Build NSX example. It serves as a great foundation to learning PowerNSX and PowerCLI including getting comfortable with PowerNSX's potential.

Chapter Summary

This chapter has introduced some tools that have been created using PowerNSX. The Visio diagramming tool provides a quick and easy way to visualize infrastructure based on data. The DFW to Excel documentation tool delivers a quick way to export firewall information and use it for offline analysis.

It is recommended administrators using PowerNSX give these tools a test in their lab and see if they are suitable for their environment.

PowerNSX provides a foundation on which numerous tools and functions can be built. The authors have heard of many great and wonderful things being created by all caliber of VMware users.

Chapter 13

Using PowerNSX to interact directly with the NSX API

Everything discussed in this booklet so far has required someone to previously conceive of the need for a given cmdlet, envisage how the pipeline operations for the related set of cmdlets might work, design how to make them function consistently with the rest of PowerNSX, and then code it. While PowerNSX has come a long way in the last two years, there are still plenty of less utilized functionality in NSX that it does not yet support *natively.*

The following information will help those who wish to extend on the native functionality of PowerNSX in their own scripts and environments. This can sometimes be trivially easy, and other times more complex depending on the task, so this area should be considered targeted at the more experienced PowerShell user.

The Core of PowerNSX

To interact with a typical XML based REST API like NSX, every call PowerNSX makes requires an HTTP request. This request includes encoded credentials in the authorization header, a content-type header, and often a properly formatted XML body.

Internally, PowerNSX uses two functions called `Invoke-NsxRestMethod` and `Invoke-NsxWebRequest` to automatically do all this. This process is based on information stored in the `Connection` object that was defined by the call to `Connect-NsxServer`, so that the calling function just needs to specify the API Uniform Resource Locator (URI), the method (e.g., put, post, get, delete), and any required body.

PowerNSX exports these cmdlets too, simplifying the process for the PowerNSX user.

Rather than having to define authorization headers as with the native PowerShell cmdlet `Invoke-RestMethod`, simply call `Connect-NsxServer`, and then `Invoke-NsxRestMethod` with the necessary `uri` and method. Example 13.1 is a simple example of this, retrieving all the services defined in globalroot-0.

NOTE
Internally, NSX refers to services as applications. This highlights another aspect of PowerNSX, removing interaction with the unfortunately frequent differences between names and concepts presented in the NSX UI and API. So PowerNSX refers to transport zones as Transport Zones, not VDN scopes, and services as Services, not applications.

Example 13.1 Using Invoke-NsxRestMethod

```
PowerCLI C:\> Connect-NsxServer -vCenterServer vc-01a

Version            : 6.3.1
BuildNumber        : 5124716
Credential         : System.Management.Automation.
PSCredential
Server             : 192.168.119.201
Port               : 443
Protocol           : https
```

```
ValidateCertificate : False
VIConnection         : winvc-01a
DebugLogging         : False
DebugLogfile        : C:\Users\Nick\AppData\Local\Temp\
PowerNSXLog-administrator@vsphere.local@-
2017_07_08_09_38_48.log


PowerCLI C:\> Invoke-NsxRestMethod -URI "/api/2.0/
services/application/scope/globalroot-0" -method get

xml                                list
---                                ----
version="1.0" encoding="UTF-8" list
```

Here `Invoke-NsxRestMethod` returns the base XML document, whereas native PowerNSX cmdlets usually return specific xml elements contained within the document, or often, collections of these XML element objects. The process to get similar output to a PowerNSX cmdlet is best explained in a couple of examples.

Retrieving Information from NSX using Invoke-NsxRestMethod

The contents of the list property is presented in Example 13.2

Example 13.2 Examining an XML Document

```
PowerCLI C:\> $xmldoc = Invoke-NsxRestMethod -URI "/
api/2.0/services/application/scope/globalroot-0" -method
get
PowerCLI C:\> $xmldoc.list

application
-----------
{IPv6-ICMP Neighbor Advertisement, edgeservice, sys-
gen-empty-app-edge-fw, IPv6-ICMP Neighbor
Solicitation...}
```

This shows the names of services within a collection of objects within a single `application` property.

This makes much more sense in the context the actual XML. One of PowerNSX's hidden gems - an internal export function that makes life easier for the casual user - `Format-Xml` - is detailed in example 13.3.

This function displays valid XML text in a manner formatted for easy consumption by a human reader

Example 13.3 Using Format-List to Display Formatted XML

```
PowerCLI C:\> $xmldoc.list | format-xml
<list>
  <application>
    <objectId>application-26</objectId>
    <objectTypeName>Application</objectTypeName>
    <vsmUuid>564D0978-80C8-D088-D259-8ED68F9852A3</
vsmUuid>
    <nodeId>02445238-635f-434f-b12e-aa8caf4c2298</nodeId>
    <revision>1</revision>
    <type>
      <typeName>Application</typeName>
    </type>
    <name>IPv6-ICMP Neighbor Advertisement</name>
    <scope>
      <id>globalroot-0</id>
      <objectTypeName>GlobalRoot</objectTypeName>
      <name>Global</name>
    </scope>
    <clientHandle></clientHandle>
    <extendedAttributes />
    <isUniversal>false</isUniversal>
    <universalRevision>0</universalRevision>
    <inheritanceAllowed>true</inheritanceAllowed>
    <element>
      <applicationProtocol>IPV6ICMP</
applicationProtocol>
      <value>neighbor-advertisement</value>
    </element>
  </application>
  <application>
    <objectId>application-1475</objectId>
    <objectTypeName>Application</objectTypeName>
    <vsmUuid>564D0978-80C8-D088-D259-8ED68F9852A3</
vsmUuid>
    <nodeId>02445238-635f-434f-b12e-aa8caf4c2298</nodeId>
    <revision>2</revision>
    <type>
      <typeName>Application</typeName>
    </type>
    <name>edgeservice</name>
    <description>localedge service</description>
    <scope>
      <id>edge-485</id>
      <objectTypeName>Edge</objectTypeName>
      <name>testedge</name>
    </scope>
```

```
<clientHandle></clientHandle>
    <extendedAttributes />
    <isUniversal>false</isUniversal>
    <universalRevision>0</universalRevision>
    <inheritanceAllowed>false</inheritanceAllowed>
    <element>
      <applicationProtocol>TCP</applicationProtocol>
      <value>1234</value>
    </element>
  </application>

...
```

This shows that a single `list` element contains multiple `application` elements. When traversing an extra level deep using dot notation and specifying the name of an element that occurs more than once, PowerShell outputs a collection of XML elements. Example 13.4 demonstrates retrieval of the collection of application XML element objects from the object returned from `Invoke-NsxrestMethod`.

Example 13.4 Retrieving a Collection of XML Elements

```
PowerCLI C:\> $xmldoc.list.application


objectId            : application-26
objectTypeName      : Application
vsmUuid             : 564D0978-80C8-D088-D259-
8ED68F9852A3
nodeId              : 02445238-635f-434f-b12e-
aa8caf4c2298
revision            : 1
type                : type
name                : IPv6-ICMP Neighbor Advertisement
scope               : scope
clientHandle        :
extendedAttributes :
isUniversal         : false
universalRevision   : 0
inheritanceAllowed : true
element             : element

objectId            : application-1475
objectTypeName      : Application
vsmUuid             : 564D0978-80C8-D088-D259-
8ED68F9852A3
nodeId              : 02445238-635f-434f-b12e-
aa8caf4c2298
```

```
revision             : 2
type                 : type
name                 : edgeservice
description          : localedge service
scope                : scope
clientHandle         :
extendedAttributes   :
isUniversal          : false
universalRevision    : 0
inheritanceAllowed   : false
element              : element

...
```

Comparing the output in this above example to that of `Get-NsxService`, it is apparent that they are the same. And like `Get-NsxService`, because it provides a collection of objects, pipeline processes are supported, allowing for iteration using `ForEach-Object`, filtering with `Where-Object`, or extracting specific properties through `Select-Object`.

Modifying Configuration using Invoke-NsxRestMethod

Extending on these examples, the NSX API documentation shows that to update a service, it is necessary to get the current definition, modify the required property, and then send the entire definition back using the method `PUT` to the URI `/2.0/services/application/{applicationId}`. Example 13.5 demonstrates how to retrieve the service `application-1476`, update the description, and then push the change back to NSX.

Example 13.5 Updating the Description of an Existing Service

```
PowerCLI C:\> $xmldoc = Invoke-NsxRestMethod -URI "/
api/2.0/services/application/application-1476" -method
get
PowerCLI C:\> $xmldoc.application


objectId             : application-1476
objectTypeName       : Application
vsmUuid              : 564D0978-80C8-D088-D259-
8ED68F9852A3
nodeId               : 02445238-635f-434f-b12e-
aa8caf4c2298
```

```
revision             : 1
type                 : type
name                 : TestService
description          : WrongDescription
scope                : scope
clientHandle         :
extendedAttributes   :
isUniversal          : false
universalRevision    : 0
inheritanceAllowed : false
element              : element


PowerCLI C:\> $xmldoc.application.description =
"RightDescription"
PowerCLI C:\> $xmldoc.application.OuterXml
<application><objectId>application-1476</objectId><obje
ctTypeName>Application</
objectTypeName><vsmUuid>564D0978-80C8-D088-D259-
8ED68F9852A3</vsmUuid><nodeId>02445238-635f-434f-b12e-
aa8caf4c2298</nodeId><revision>1</revision><type><typeN
ame>Application</typeName></type><name>TestService</
name><description>RightDescription</description><scope
><id>globalroot-0</id><objectTypeName>GlobalRoot</
objectTypeName><name>Global</name></
scope><clientHandle></clientHandle><extendedAttributes
/><isUniversal>false</isUniversal><universalRevis
ion>0</universalRevision><inheritanceAllowed>false</
inheritanceAllowed><element><applicationProtocol>TCP</
applicationProtocol><value>1234</value></element></
application>
PowerCLI C:\> Invoke-NsxRestMethod -method put -URI "/
api/2.0/services/application/application-1476" -body
$xmldoc.application.
OuterXml

xml                                 application
---                                 -----------
version="1.0" encoding="UTF-8" application
```

Why is PowerNSX required at all? While updating simple properties - technically existing xml text elements within the xml definition of existing NSX objects - is easy and frequently useful, that is about where the simplicity ends.

Things start getting much harder when faced with the need to create new, non-text elements, usually with one or more sub-elements, deal with xml attributes, or efficiently search for specific elements within many elements.

The realities associated with direct manipulation of XML quickly become apparent. While the .NET classes for handling XML are very capable, XML can be quite complex to manipulate effectively and efficiently. The techniques are more developer oriented than they are PowerShell-focused and will be alien to the average administrator.

For this reason, PowerNSX must do a significant amount of work to streamline XML tasks.

NOTE
More advanced XML manipulation is beyond the scope of this book. For those with skills in that area, please consider contributing to PowerNSX!

About Invoke-NsxWebRequest

The introduction to this chapter mentioned `Invoke-NsxWebRequest`. Like the native PowerShell cmdlet `Invoke-RestMethod` that it is based on, `Invoke-NsxRestMethod` returns a native PowerShell object representing the response content. This is essentially the *easy button* - eliminating explicit conversion of the response content to xml. It provides an XML object back ready for direct use.

There are limitations to the `Invoke-NsxRestMethod` cmdlet, primarily lack of access the response headers. Many times, this is not an issue, but there are many parts of the NSX API that return data where access to the response headers is required. It is then time to use `Invoke-NsxWebRequest`. There are other reasons, like wanting to access the response content as raw text or access the response code, and these are all available as properties of the `WebResponse` object returned by `Invoke-NsxWebRequest`.

Like the `Invoke-NsxRestMethod` cmdlet, the `Invoke-NsxWebRequest` cmdlet also absolves the need to populate authorization or content-type headers, using the `$Connection` variable to automatically populate these. If access is required the response headers of an API call to NSX, then `Invoke-NsxWebRequest` is still useful, although examples of this are beyond the scope of this book.

Chapter Summary

While delving into the innards of the NSX API more than the average PowerNSX user is required to, using the core `Invoke-NsxRestMethod` and `Invoke-NsxWebRequest` cmdlet allows very easy interaction with *any* part of the NSX API required in the same PowerNSX script.

The examples in this chapter demonstrate that `Invoke-NsxRestMethod` just returns an XML object, precisely as do many PowerNSX cmdlets. Often, a PowerNSX cmdlet is doing little more than constructing the body XML based on passed parameters, calling `Invoke-NsxRestMethod`, and returning XML elements contained within the response XML document.

This also demonstrates one of PowerShell's many strengths that come from its object-oriented behavior; it automatically does a very, very good job of making XML behave like a native PowerShell object, and PowerNSX relies heavily on that fact! It is primarily for this reason - that even if PowerNSX does not have native functionality - PowerShell remains a very powerful choice of platform for providing a CLI and automation capability for NSX.

Appendix

Acronyms

ANZ	Australia / New Zealand
API	Application Programmable Interface
BGP	Border Gateway Protocol
CLI	Command Line Interface
Cmdlet	Command-let
DFW	Distributed Firewall
DLR	Distributed Logical Router
ESG	NSX Edge Services Gateway
HTTP	Hyper Text Transfer Protocol
HTTPS	Hyper Text Transfer Protocol Secure
OSPF	Open Shortest Path First
LIF	Logical Interface
REST	Representative State Transfer
TCP	Transmission Control Protocol
UDP	User Datagram Protocol
VIP	Virtual IP Address
VXLAN	Virtual Extensible Local Area Network
XML	eXtensible Markup Language

Index